Teaching Resources

Grade 2

Harcourt Brace & Company

Orlando • Atlanta • Austin • Boston • San Francisco • Chicago • Dallas • New York • Toronto • London

http://www.hbschool.com

Printed in the United States of America

ISBN 0-15 311117-8

6 7 8 9 10 085 2001

CONTENTS

TEACHING RESOURCES

Various types of resources for lessons and practice activities in the Teacher's Edition are included in this section. Some of these resources may also be used with the Learning Center Cards for this grade level.

The resources are provided for the following categories:

▶ **Number and Operations**
▶ **Money**
▶ **Time**
▶ **Measurement**
▶ **Geometry**
▶ **Data and Graphing**
▶ **Workmats**
▶ **Other**

$$\begin{array}{r} 0 \\ +1 \\ \hline \end{array}$$ $$\begin{array}{r} 0 \\ +2 \\ \hline \end{array}$$ $$\begin{array}{r} 0 \\ +3 \\ \hline \end{array}$$

$$\begin{array}{r} 0 \\ +4 \\ \hline \end{array}$$ $$\begin{array}{r} 0 \\ +5 \\ \hline \end{array}$$ $$\begin{array}{r} 0 \\ +6 \\ \hline \end{array}$$

$$\begin{array}{r} 0 \\ +7 \\ \hline \end{array}$$ $$\begin{array}{r} 0 \\ +8 \\ \hline \end{array}$$ $$\begin{array}{r} 0 \\ +9 \\ \hline \end{array}$$

1 +0	1 +1	1 +2
1 +3	1 +4	1 +5
1 +6	1 +7	1 +8

2 +0	2 +1	2 +2
2 +3	2 +4	2 +5
2 +6	2 +7	3 +0

3 +1	3 +2	3 +3
3 +4	3 +5	3 +6
4 +0	4 +1	4 +2

$$\begin{array}{r} 4 \\ +3 \\ \hline \end{array}$$

$$\begin{array}{r} 4 \\ +4 \\ \hline \end{array}$$

$$\begin{array}{r} 4 \\ +5 \\ \hline \end{array}$$

$$\begin{array}{r} 5 \\ +0 \\ \hline \end{array}$$

$$\begin{array}{r} 5 \\ +1 \\ \hline \end{array}$$

$$\begin{array}{r} 5 \\ +2 \\ \hline \end{array}$$

$$\begin{array}{r} 5 \\ +3 \\ \hline \end{array}$$

$$\begin{array}{r} 5 \\ +4 \\ \hline \end{array}$$

$$\begin{array}{r} 6 \\ +0 \\ \hline \end{array}$$

Addition Fact Cards

6 +1	6 +2	6 +3
7 +0	7 +1	7 +2
8 +0	8 +1	9 +0

Addition Fact Cards

$$\begin{array}{r} 1 \\ +9 \\ \hline \end{array}$$

$$\begin{array}{r} 2 \\ +8 \\ \hline \end{array}$$

$$\begin{array}{r} 2 \\ +9 \\ \hline \end{array}$$

$$\begin{array}{r} 3 \\ +7 \\ \hline \end{array}$$

$$\begin{array}{r} 3 \\ +8 \\ \hline \end{array}$$

$$\begin{array}{r} 3 \\ +9 \\ \hline \end{array}$$

$$\begin{array}{r} 4 \\ +6 \\ \hline \end{array}$$

$$\begin{array}{r} 4 \\ +7 \\ \hline \end{array}$$

$$\begin{array}{r} 4 \\ +8 \\ \hline \end{array}$$

4 +9	5 +5	5 +6
5 +7	5 +8	5 +9
6 +4	6 +5	6 +6

6 +7	6 +8	6 +9
7 +3	7 +4	7 +5
7 +6	7 +7	7 +8

$\begin{array}{r} 7 \\ +9 \\ \hline \end{array}$	$\begin{array}{r} 8 \\ +2 \\ \hline \end{array}$	$\begin{array}{r} 8 \\ +3 \\ \hline \end{array}$
$\begin{array}{r} 8 \\ +4 \\ \hline \end{array}$	$\begin{array}{r} 8 \\ +5 \\ \hline \end{array}$	$\begin{array}{r} 8 \\ +6 \\ \hline \end{array}$
$\begin{array}{r} 8 \\ +7 \\ \hline \end{array}$	$\begin{array}{r} 8 \\ +8 \\ \hline \end{array}$	$\begin{array}{r} 8 \\ +9 \\ \hline \end{array}$

$$
\begin{array}{r} 9 \\ +1 \\ \hline \end{array}
\qquad
\begin{array}{r} 9 \\ +2 \\ \hline \end{array}
\qquad
\begin{array}{r} 9 \\ +3 \\ \hline \end{array}
$$

$$
\begin{array}{r} 9 \\ +4 \\ \hline \end{array}
\qquad
\begin{array}{r} 9 \\ +5 \\ \hline \end{array}
\qquad
\begin{array}{r} 9 \\ +6 \\ \hline \end{array}
$$

$$
\begin{array}{r} 9 \\ +7 \\ \hline \end{array}
\qquad
\begin{array}{r} 9 \\ +8 \\ \hline \end{array}
\qquad
\begin{array}{r} 9 \\ +9 \\ \hline \end{array}
$$

R12

Addition Table

Base-Ten Materials (tens, ones)

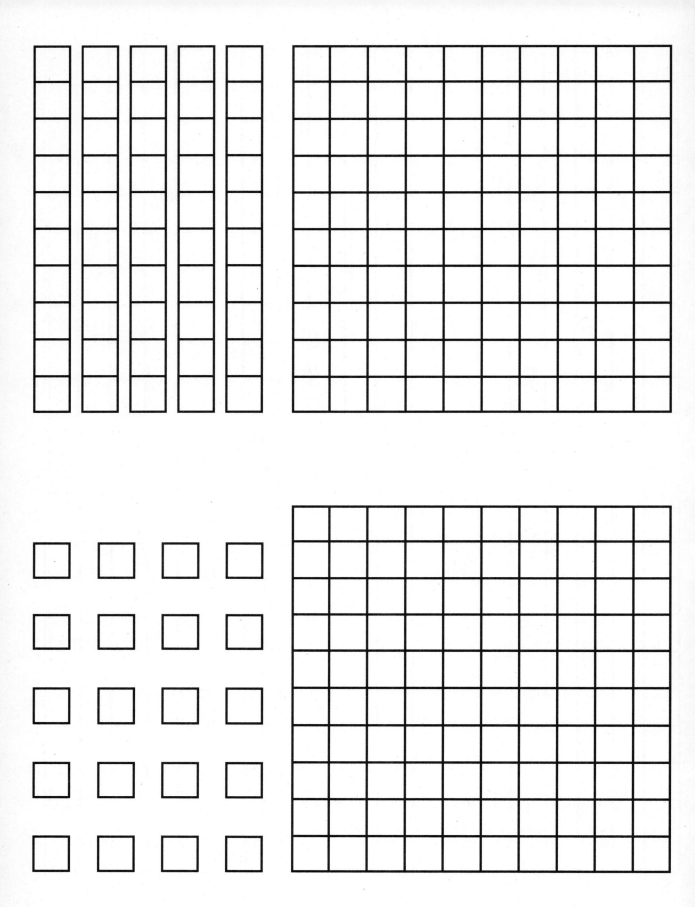

Base-Ten Materials (hundreds, tens, ones)

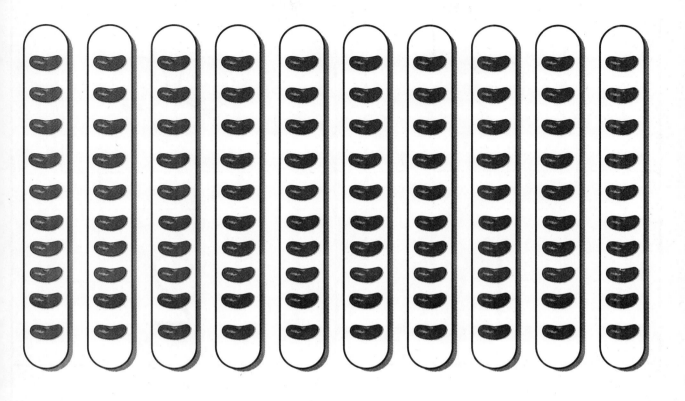

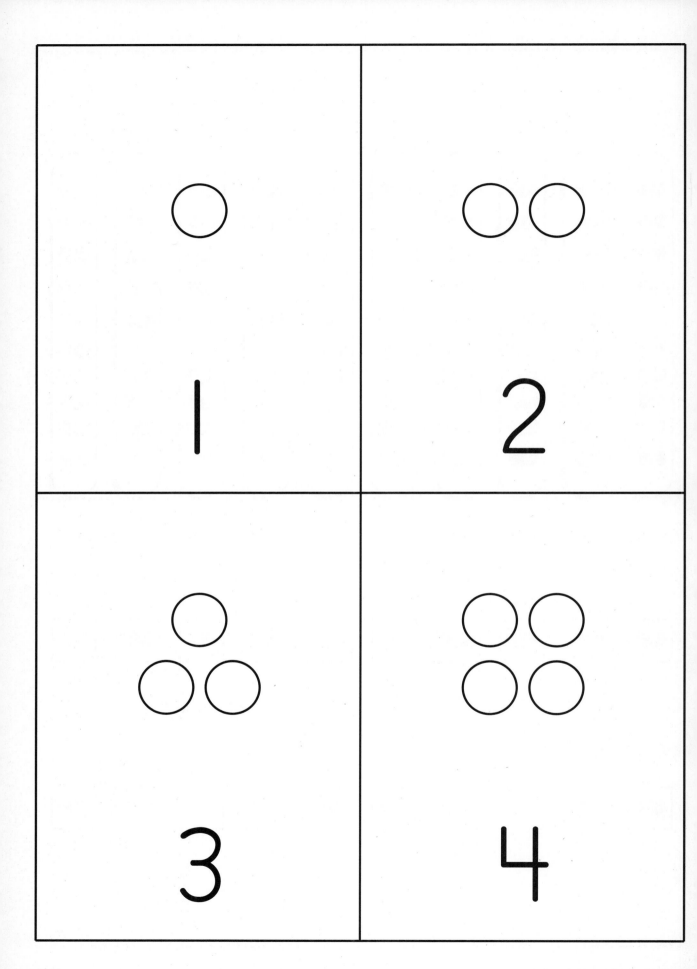

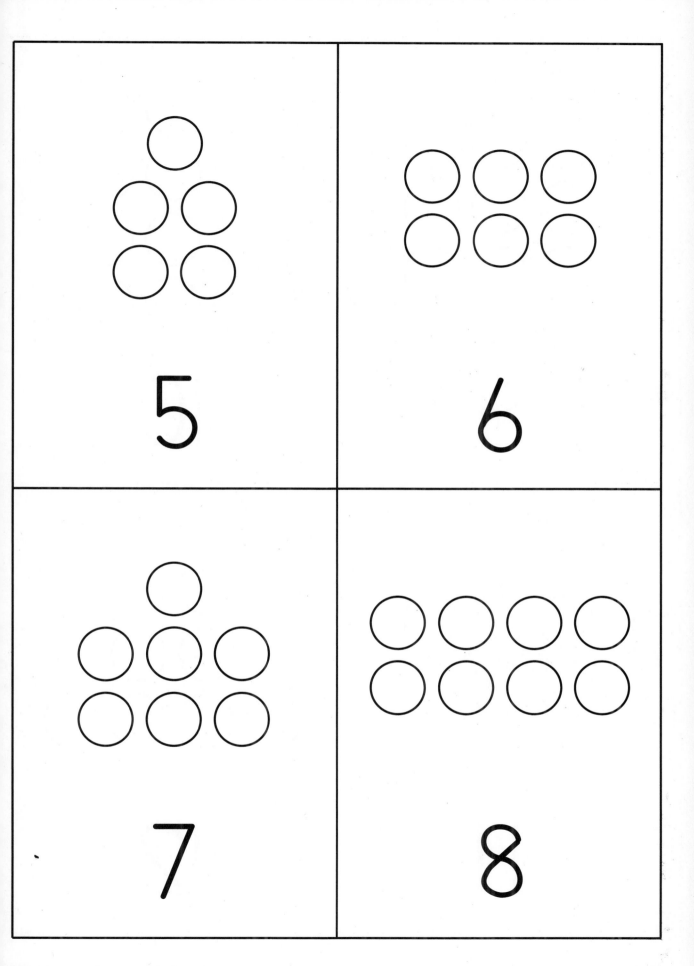

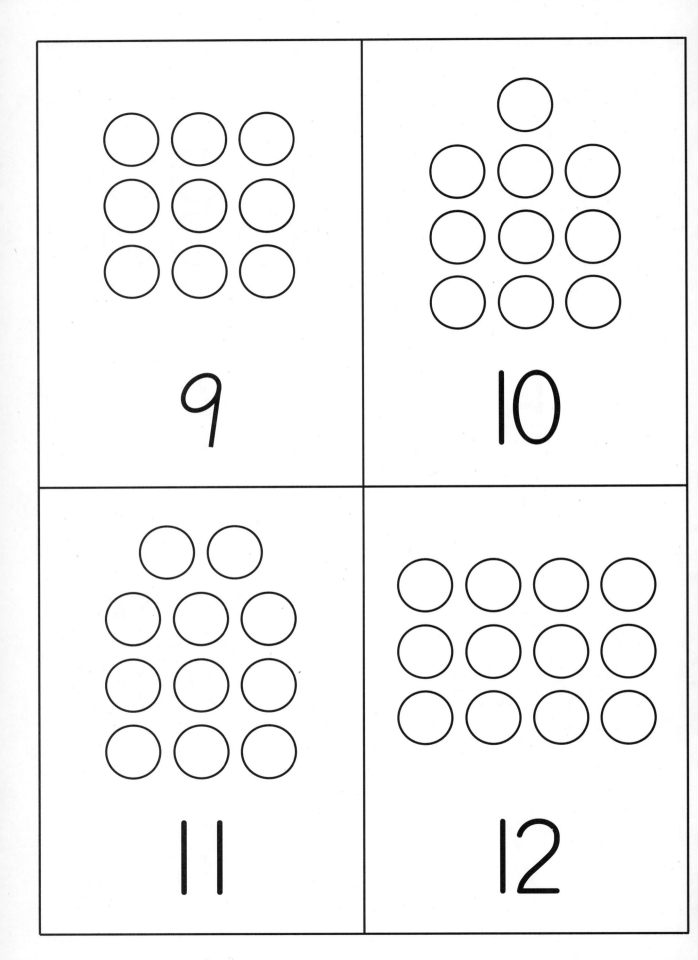

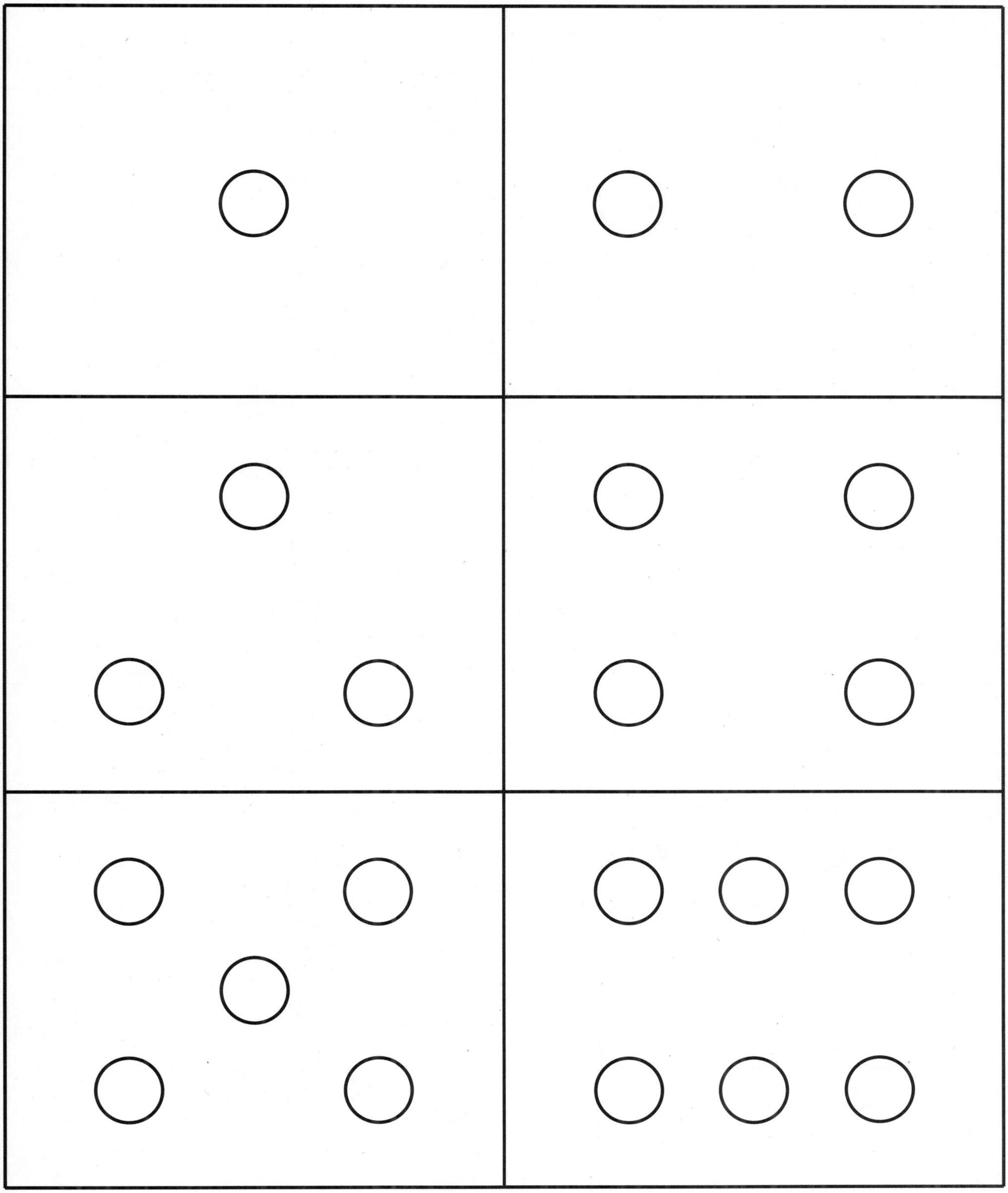

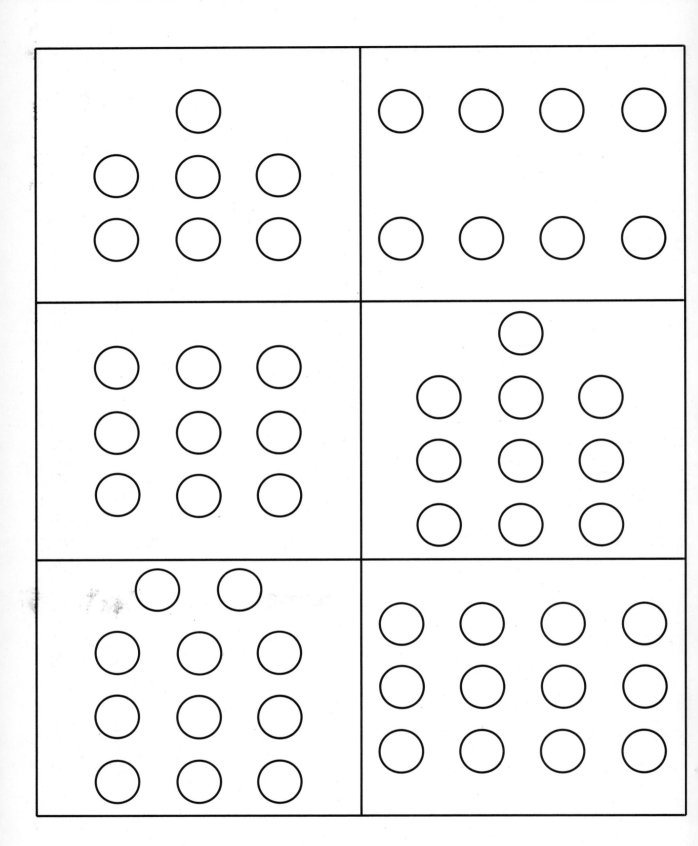

Dot Cards (7–12)

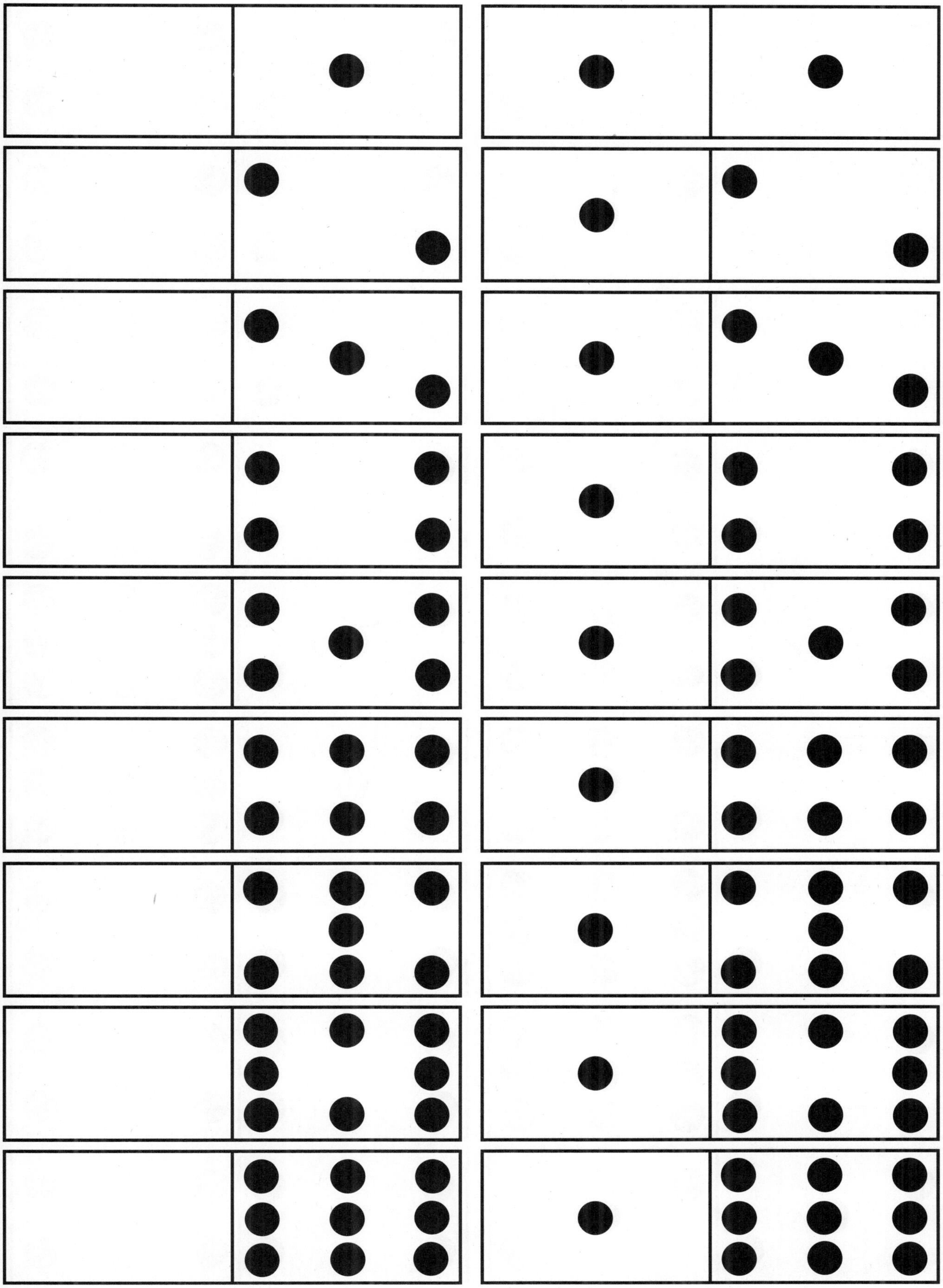

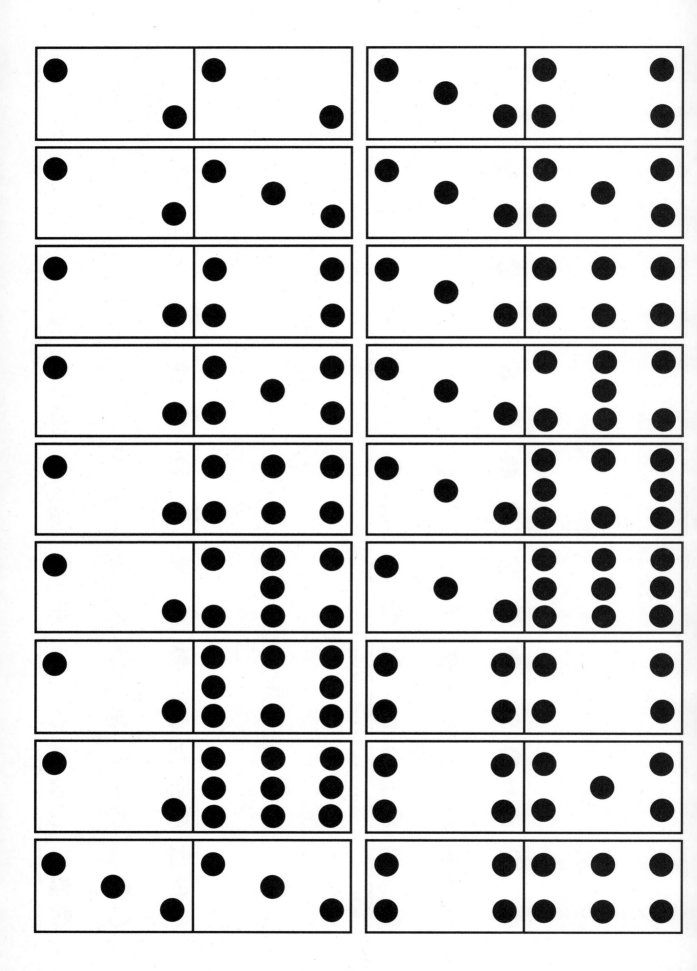

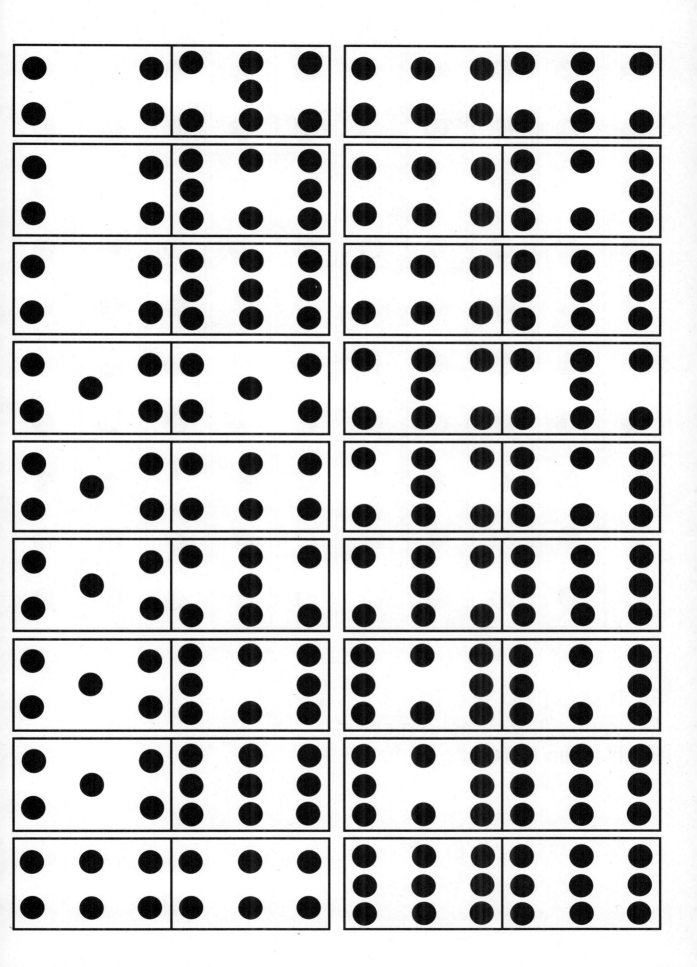

Dot Patterns (3)   R23

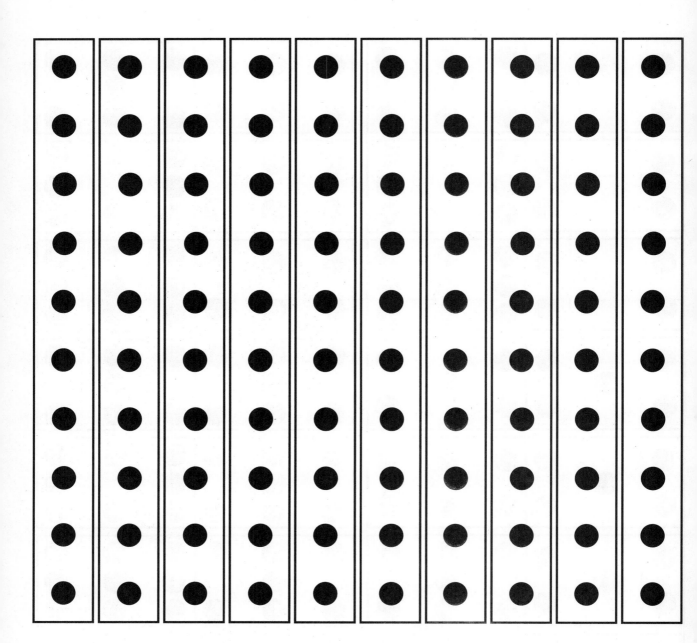

Each dot is a 1.
Each strip (link) is a 10.
Ten strips (links) make a chain.
Each chain is 100.

1	2	3	4	5	6	7	8	9	10
11	12	13	14	15	16	17	18	19	20
21	22	23	24	25	26	27	28	29	30
31	32	33	34	35	36	37	38	39	40
41	42	43	44	45	46	47	48	49	50
51	52	53	54	55	56	57	58	59	60
61	62	63	64	65	66	67	68	69	70
71	72	73	74	75	76	77	78	79	80
81	82	83	84	85	86	87	88	89	90
91	92	93	94	95	96	97	98	99	100

zero	one
two	three
four	five

six	seven
eight	nine
ten	twenty

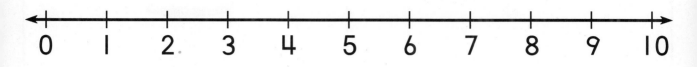

0 1 2 3 4 5 6 7 8 9 10

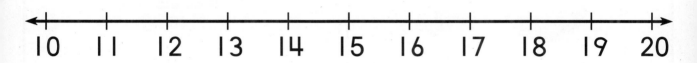

10 11 12 13 14 15 16 17 18 19 20

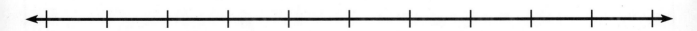

4	0
5	1
6	2
7	3

12	8
13	9
14	10
15	11

20 21 22 23

16 17 18 19

28	24
29	25
30	26
31	27

32

33

34

35

36

37

38

39

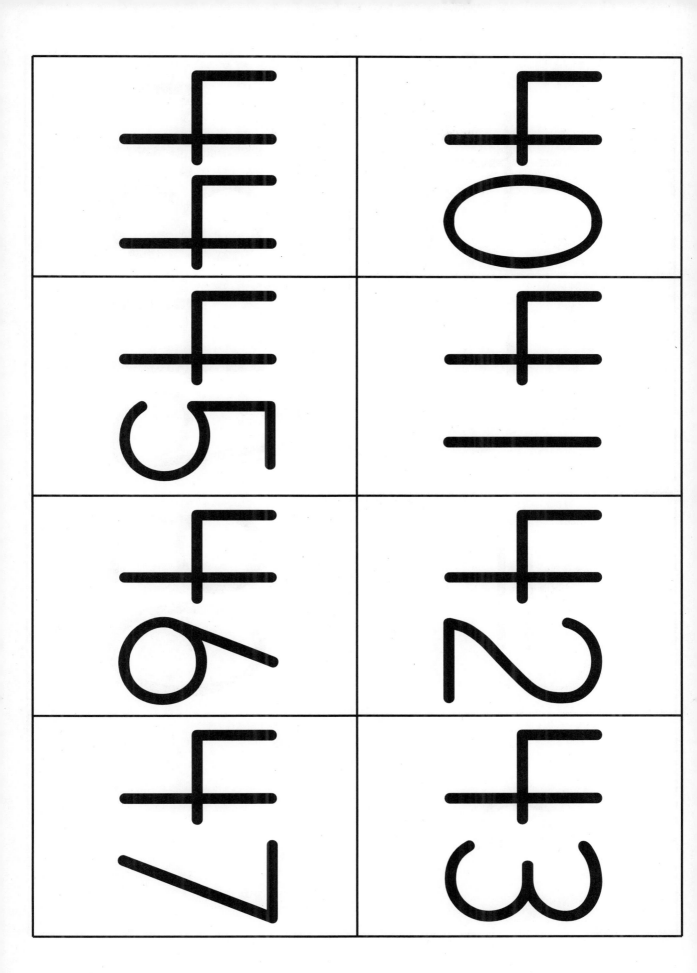

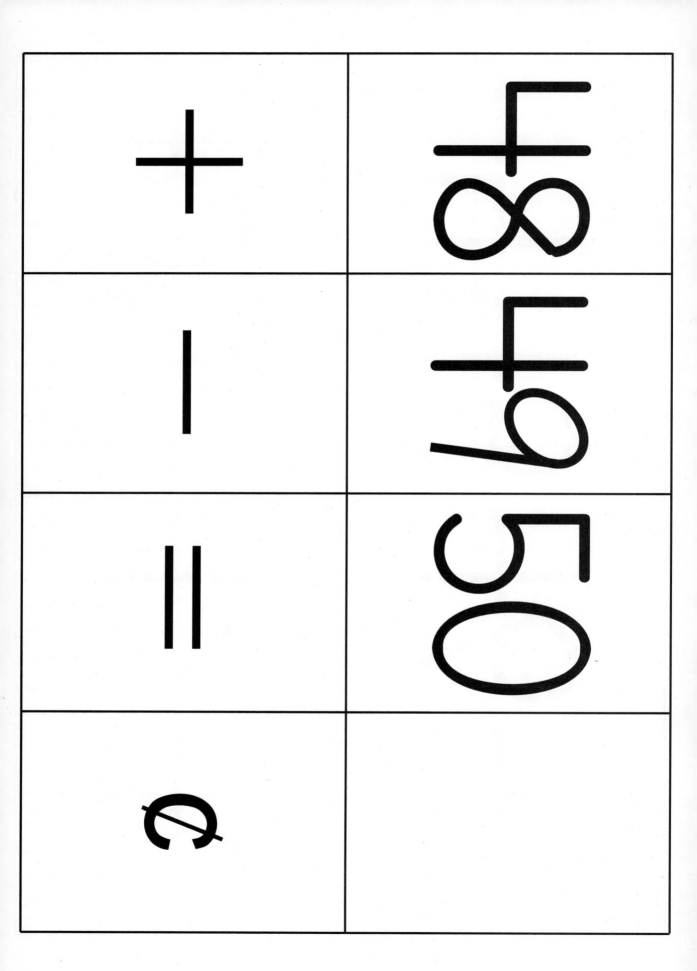

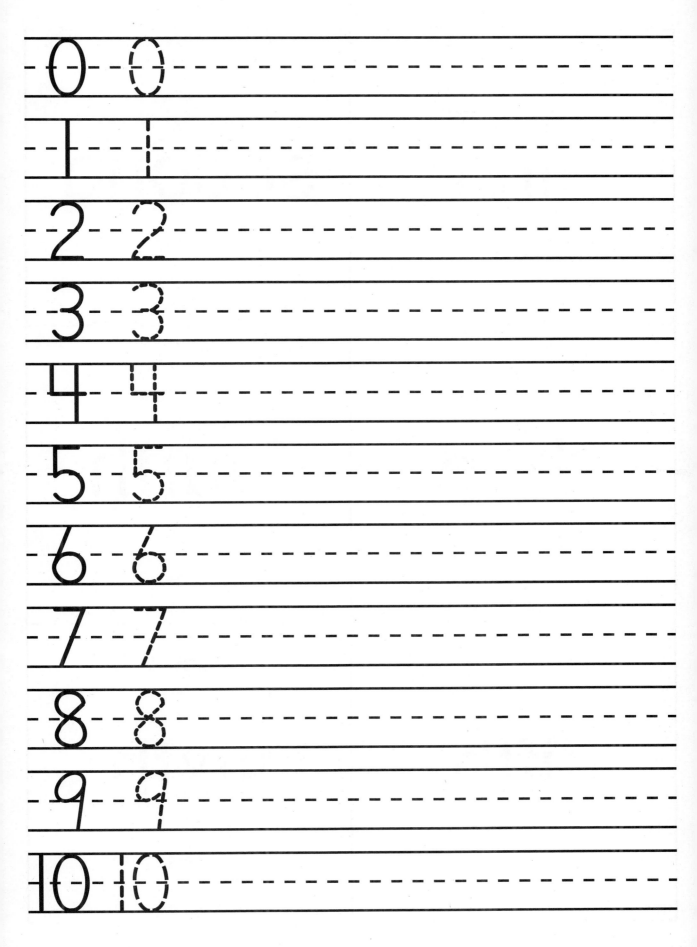

first	second
third	fourth
fifth	sixth

Ordinal Number Word Cards (first through sixth)

seventh	eighth
ninth	tenth
eleventh	twelfth

Ordinal Number Word Cards (seventh through twelfth)

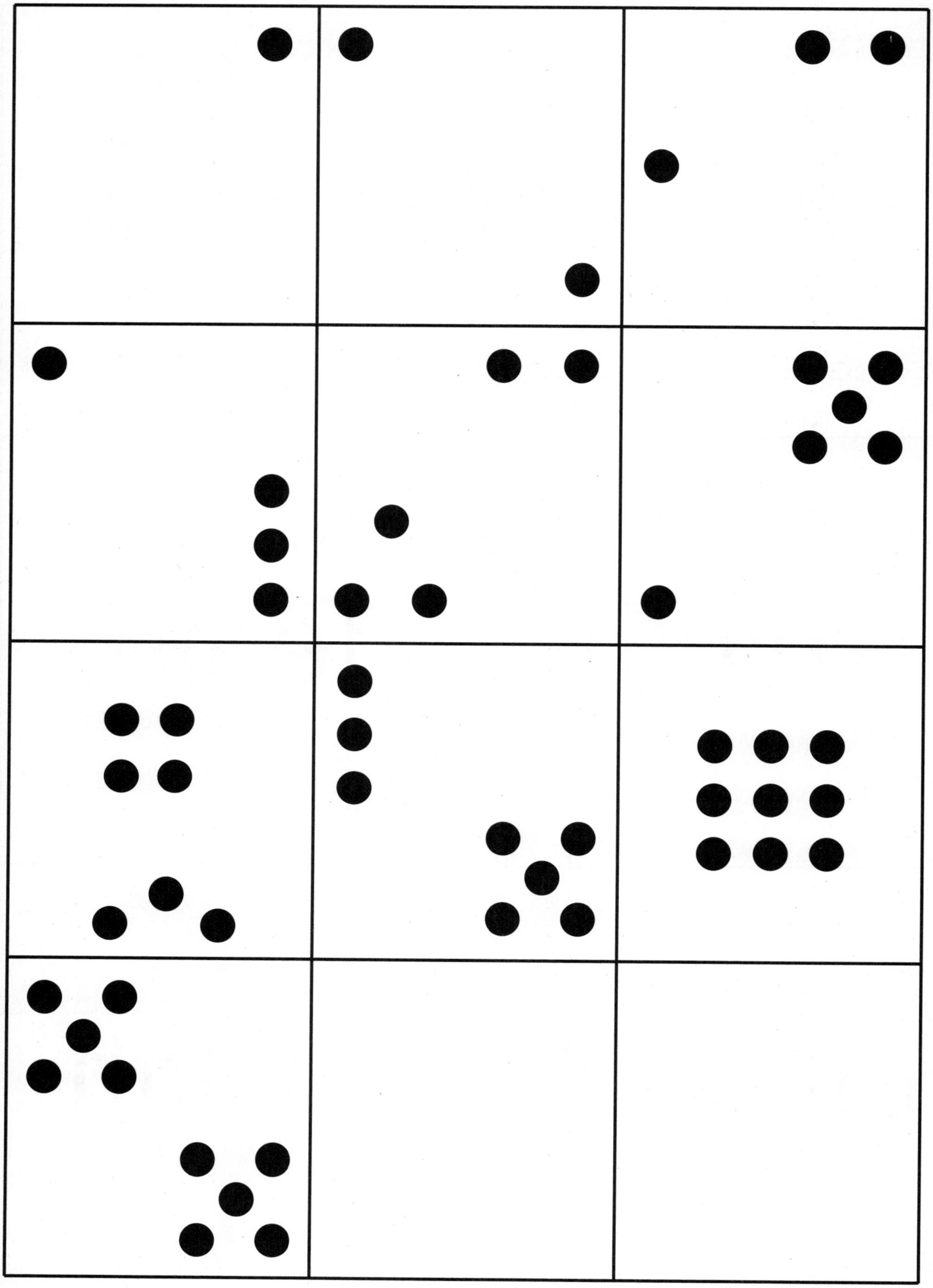

tens	ones

tens	ones

tens	ones

tens	ones

tens	ones

tens	ones

tens	ones

tens	ones

tens	ones

tens	ones

tens	ones

tens	ones

hundreds	tens	ones

hundreds	tens	ones

hundreds	tens	ones

hundreds	tens	ones

hundreds	tens	ones

hundreds	tens	ones

hundreds	tens	ones

hundreds	tens	ones

Recording Charts (hundreds, tens, ones)

| 9 | 9 | 9 |
-9	-8	-7

| 9 | 9 | 9 |
-6	-5	-4

| 9 | 9 | 9 |
-3	-2	-1

$\begin{array}{r} 9 \\ -0 \\ \hline \end{array}$	$\begin{array}{r} 8 \\ -8 \\ \hline \end{array}$	$\begin{array}{r} 8 \\ -7 \\ \hline \end{array}$
$\begin{array}{r} 8 \\ -6 \\ \hline \end{array}$	$\begin{array}{r} 8 \\ -5 \\ \hline \end{array}$	$\begin{array}{r} 8 \\ -4 \\ \hline \end{array}$
$\begin{array}{r} 8 \\ -3 \\ \hline \end{array}$	$\begin{array}{r} 8 \\ -2 \\ \hline \end{array}$	$\begin{array}{r} 8 \\ -1 \\ \hline \end{array}$

Subtraction Fact Cards

8 − 0 ―――	7 − 7 ―――	7 − 6 ―――
7 − 5 ―――	7 − 4 ―――	7 − 3 ―――
7 − 2 ―――	7 − 1 ―――	7 − 0 ―――

6 − 6 ─────	6 − 5 ─────	6 − 4 ─────
6 − 3 ─────	6 − 2 ─────	6 − 1 ─────
6 − 0 ─────	5 − 5 ─────	5 − 4 ─────

Subtraction Fact Cards

$\begin{array}{r} 5 \\ -3 \\ \hline \end{array}$	$\begin{array}{r} 5 \\ -2 \\ \hline \end{array}$	$\begin{array}{r} 5 \\ -1 \\ \hline \end{array}$
$\begin{array}{r} 5 \\ -0 \\ \hline \end{array}$	$\begin{array}{r} 4 \\ -4 \\ \hline \end{array}$	$\begin{array}{r} 4 \\ -3 \\ \hline \end{array}$
$\begin{array}{r} 4 \\ -2 \\ \hline \end{array}$	$\begin{array}{r} 4 \\ -1 \\ \hline \end{array}$	$\begin{array}{r} 4 \\ -0 \\ \hline \end{array}$

3 −3 —	3 −2 —	3 −1 —
3 −0 —	2 −2 —	2 −1 —
2 −0 —	1 −1 —	1 −0 —

Subtraction Fact Cards

$$\begin{array}{r} 10 \\ -1 \\ \hline \end{array}$$

$$\begin{array}{r} 10 \\ -2 \\ \hline \end{array}$$

$$\begin{array}{r} 10 \\ -3 \\ \hline \end{array}$$

$$\begin{array}{r} 10 \\ -4 \\ \hline \end{array}$$

$$\begin{array}{r} 10 \\ -5 \\ \hline \end{array}$$

$$\begin{array}{r} 10 \\ -6 \\ \hline \end{array}$$

$$\begin{array}{r} 10 \\ -7 \\ \hline \end{array}$$

$$\begin{array}{r} 10 \\ -8 \\ \hline \end{array}$$

$$\begin{array}{r} 10 \\ -9 \\ \hline \end{array}$$

1 1	1 1	1 1
− 2	− 3	− 4
1 1	1 1	1 1
− 5	− 6	− 7
1 1	1 1	1 2
− 8	− 9	− 3

12 −4 ——	12 −5 ——	12 −6 ——
12 −7 ——	12 −8 ——	12 −9 ——
13 −4 ——	13 −5 ——	13 −6 ——

13	13	13
− 7	− 8	− 9

14	14	14
− 5	− 6	− 7

14	14	15
− 8	− 9	− 6

Subtraction Fact Cards

15 − 7	15 − 8	15 − 9
16 − 7	16 − 8	16 − 9
17 − 8	17 − 9	18 − 9

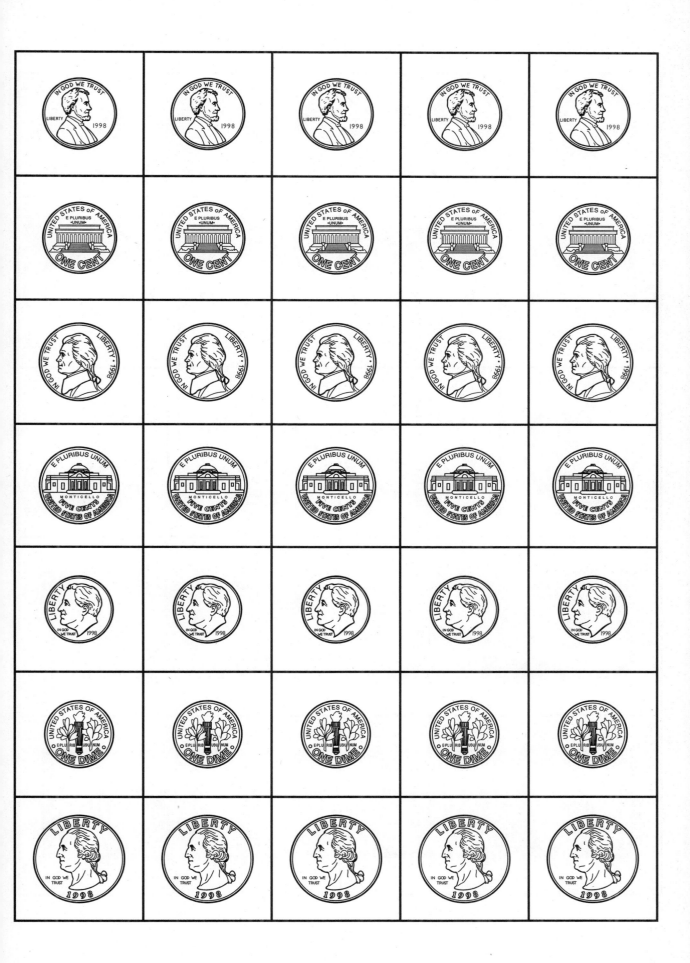

Coins

Coins and Bills

Game Cards (with price tags)

Quarter	Dime	Nickel	Penny

Money Recording Sheet

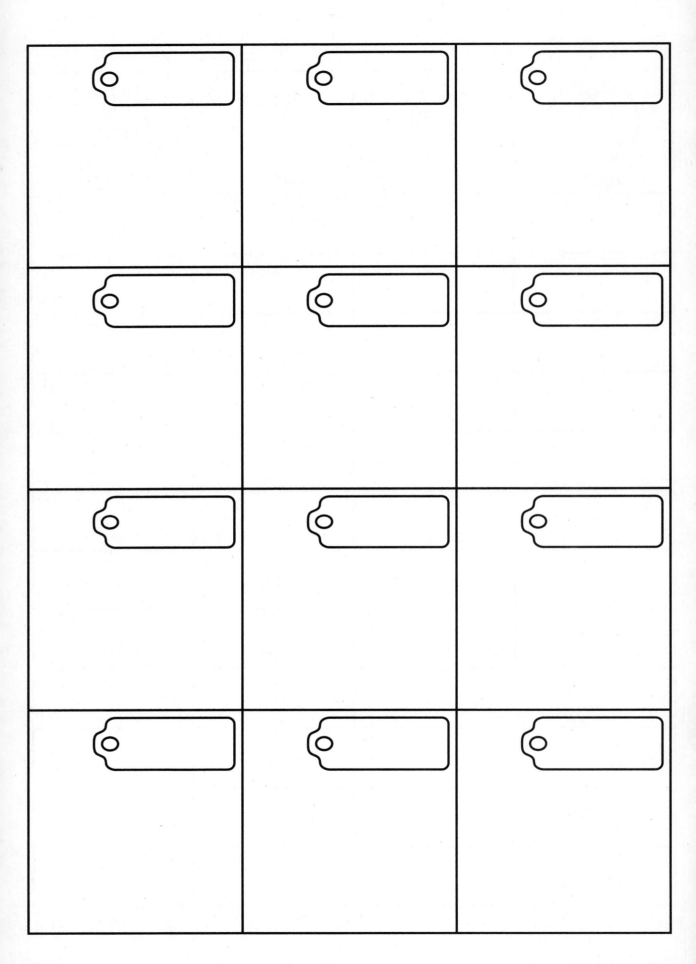

Price Tags (blank)

Toy Pictures (with price tags)

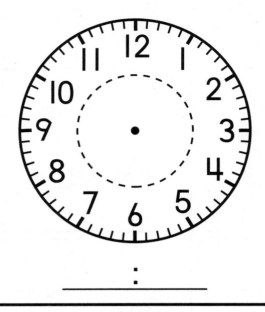

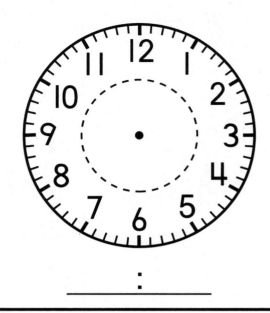

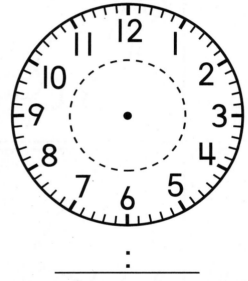

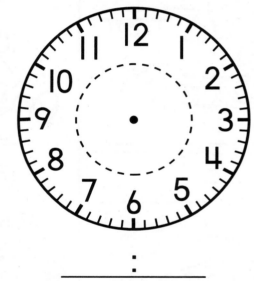

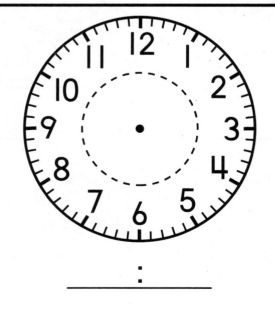

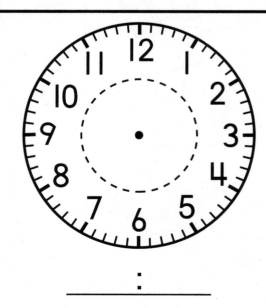

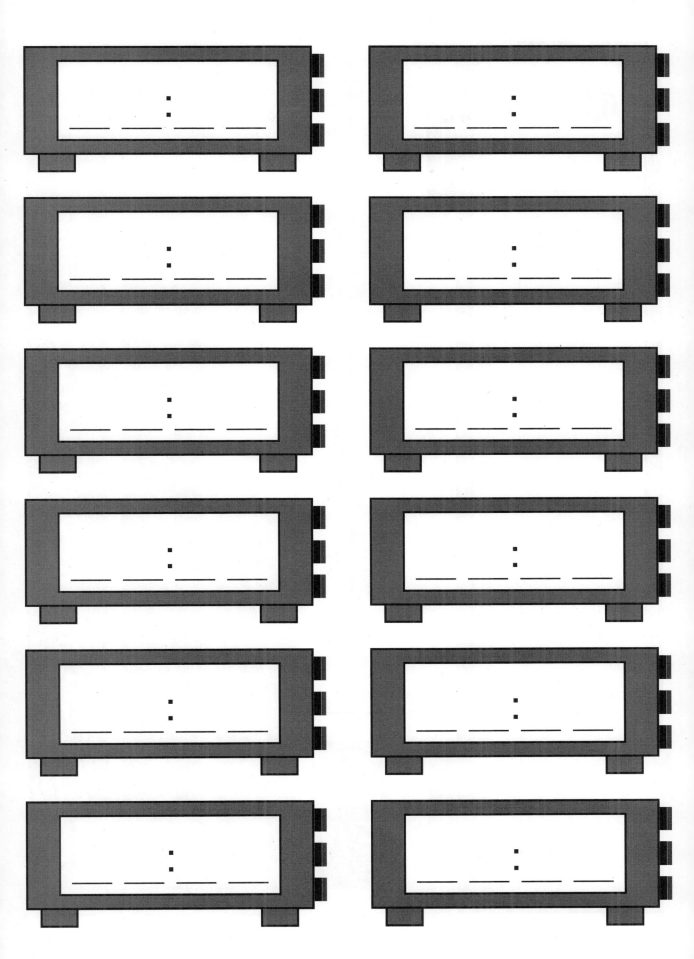

Digital Clockfaces

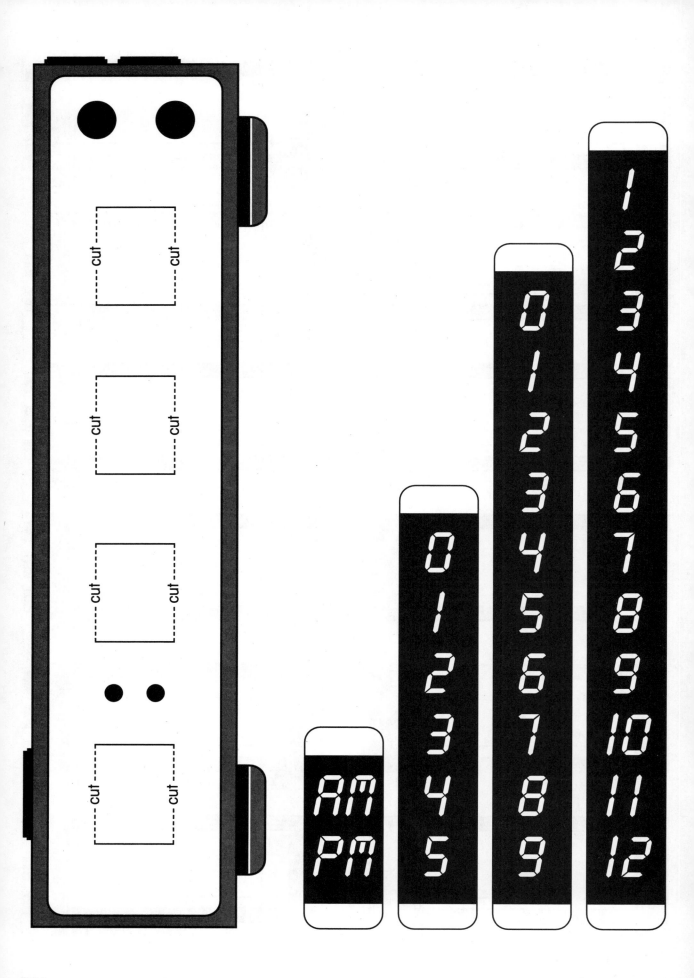

Digital Clock Model

Sunday	Monday	Tuesday	Wednesday	Thursday	Friday	Saturday

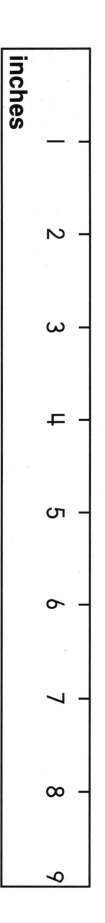

inches

1 2 3 4 5 6 7 8 9

centimeters

1 2 3 4 5 6 7 8 9 10 11 12 13 14 15 16 17 18 19 20 21 22

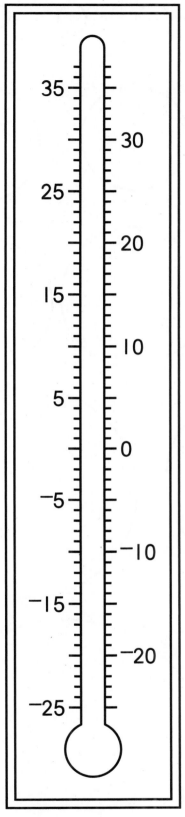

Celsius

_____°C

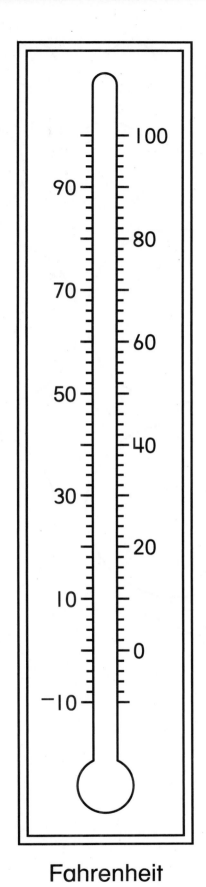

Fahrenheit

_____°F

Circles

Cone Pattern

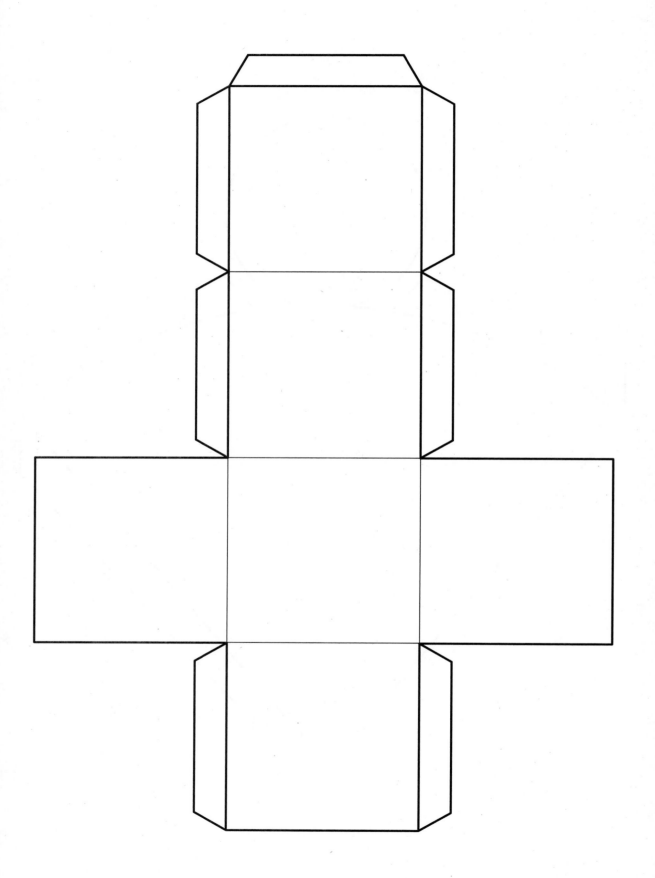

Cube Pattern

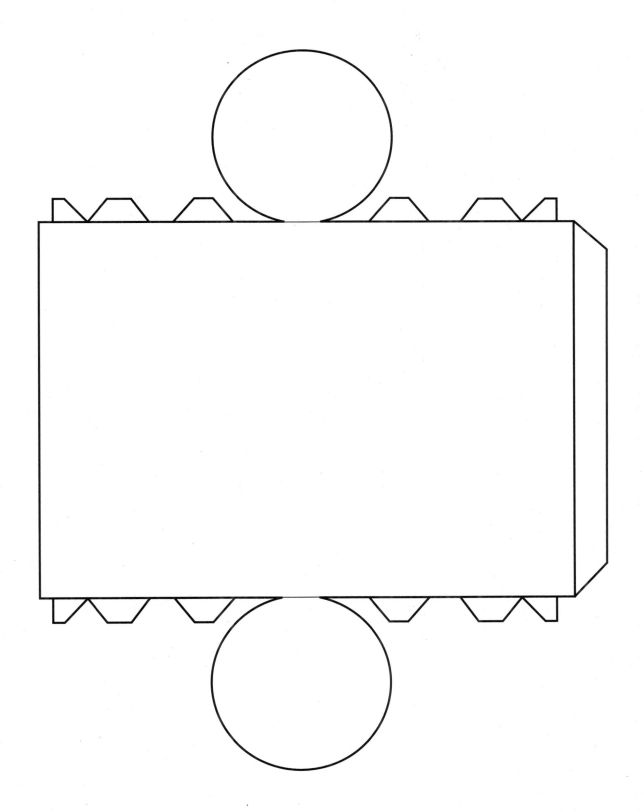

Cylinder Pattern

R74

Dot Paper

Dot Paper (isometric)

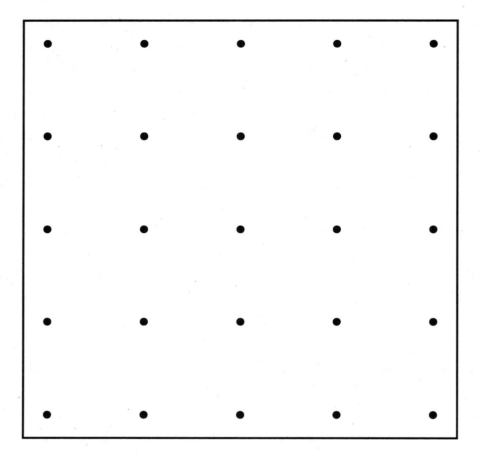

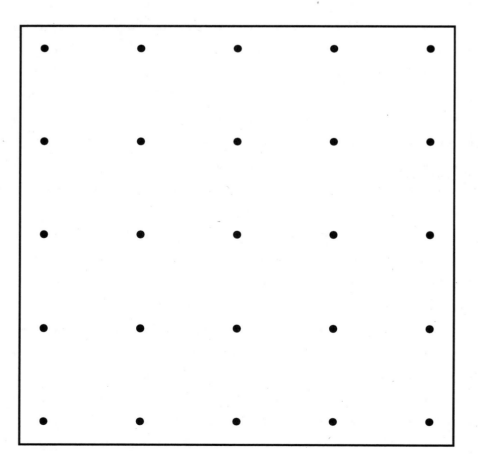

Geoboard Dot Paper

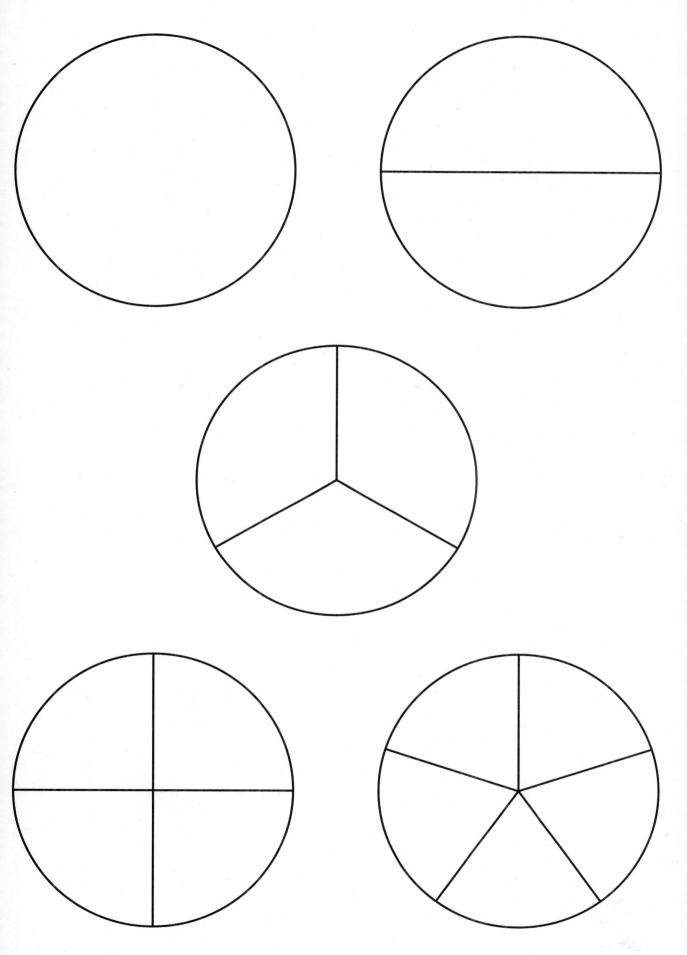

Fraction Circles (whole to fifths)

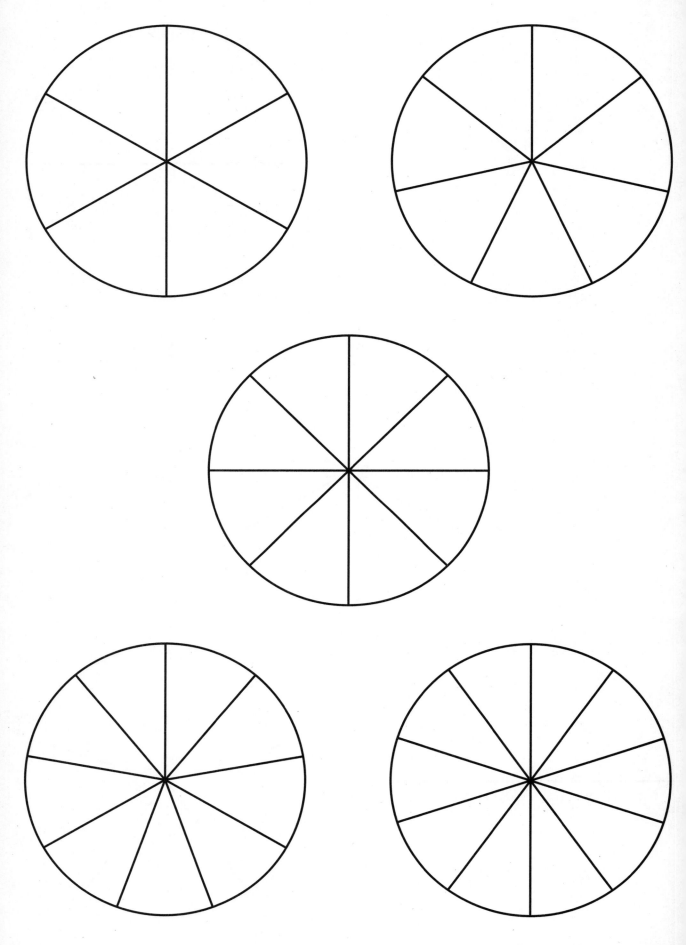

Fraction Circles (sixths to tenths)

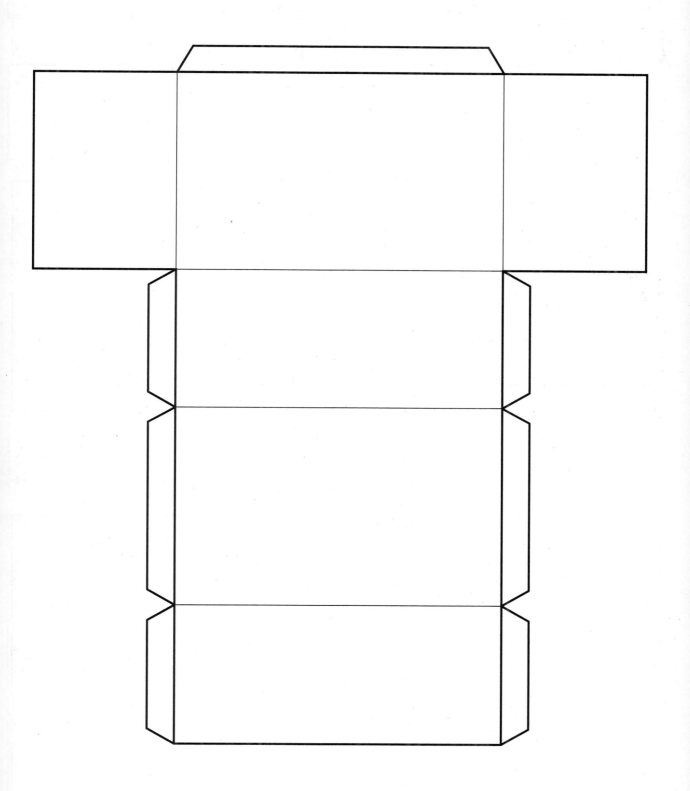

Rectangular Prism Pattern

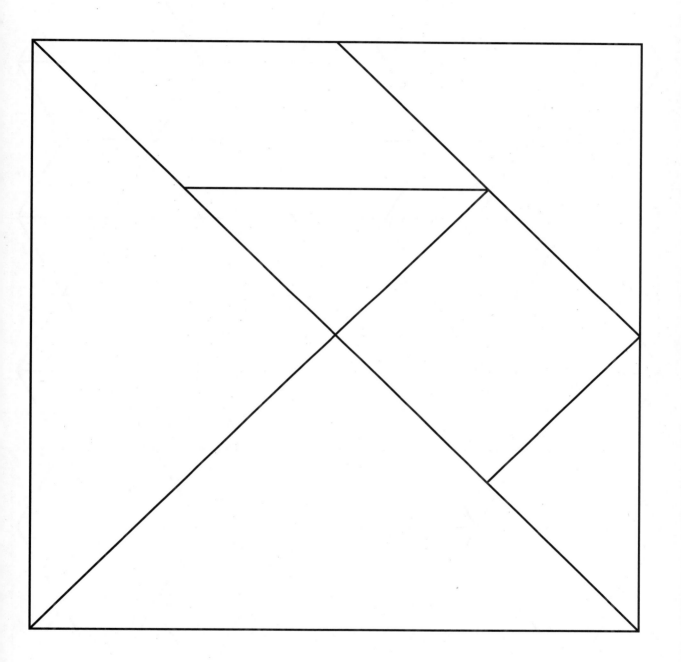

Tangram Pattern

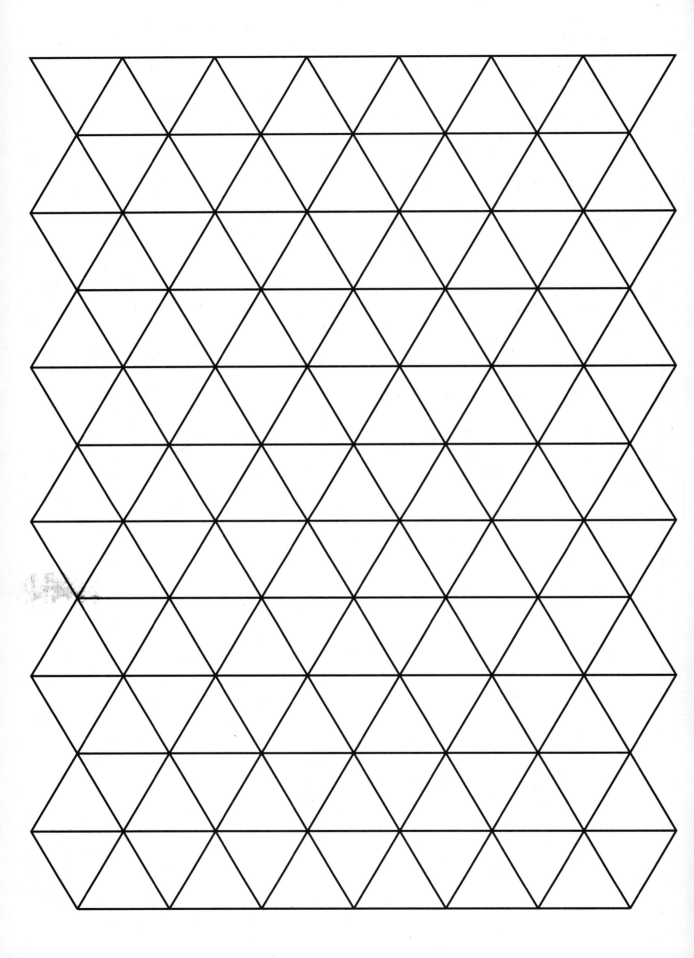

Triangle-Shaped Grid

Triangles

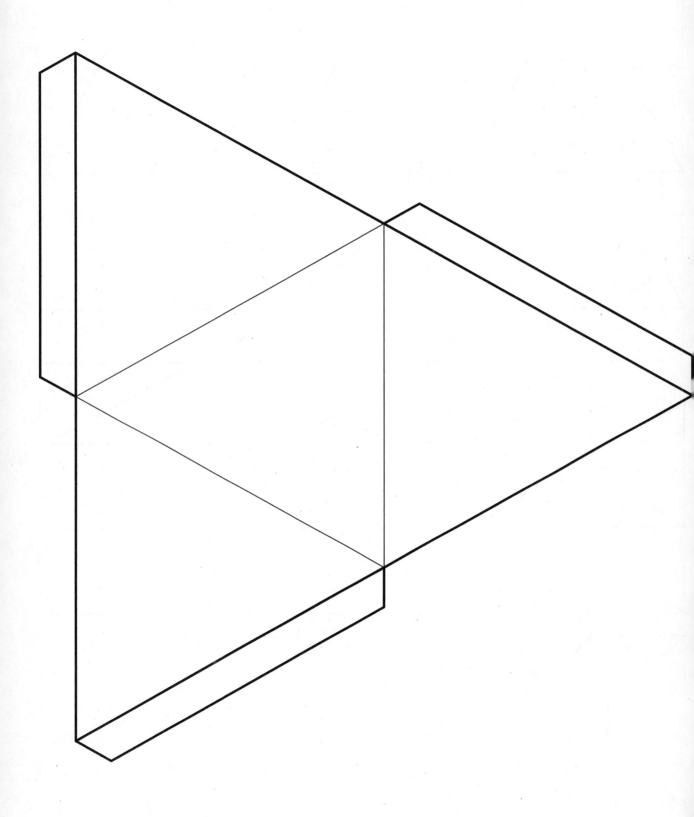

Triangular Pyramid Pattern

1-Centimeter Grid Paper

R88

1-Inch Grid Paper

Bar Graph Grid

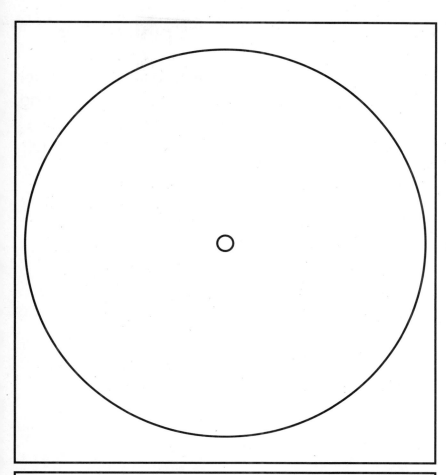

Spinner Tips

How to assemble spinner.
- Glue patterns to oaktag.
- Cut out and attach pointer with a fastener.

Alternative
- Children can use a paper clip and pencil instead.

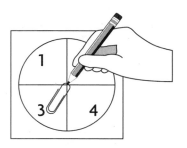

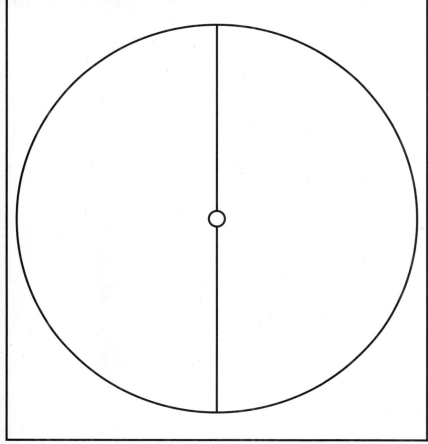

Spinners (blank and 2-section)

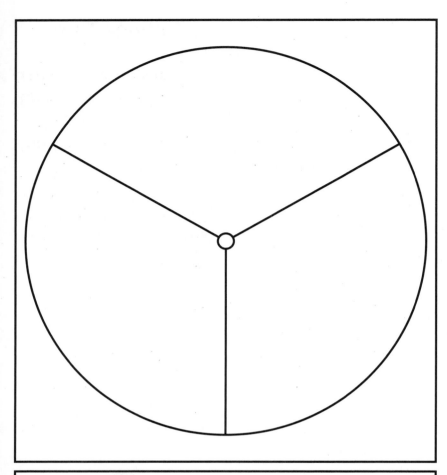

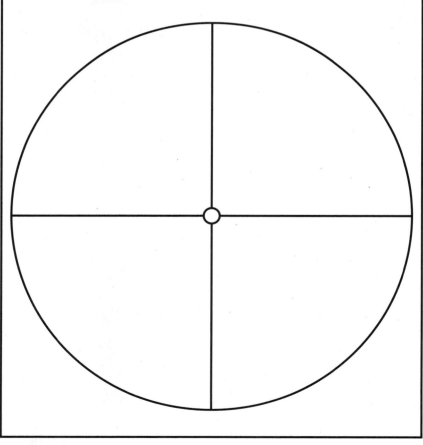

Spinner Tips

How to assemble spinner.
- Glue patterns to oaktag.
- Cut out and attach pointer with a fastener.

Alternative
- Children can use a paper clip and pencil instead.

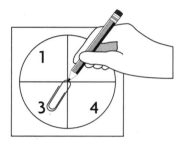

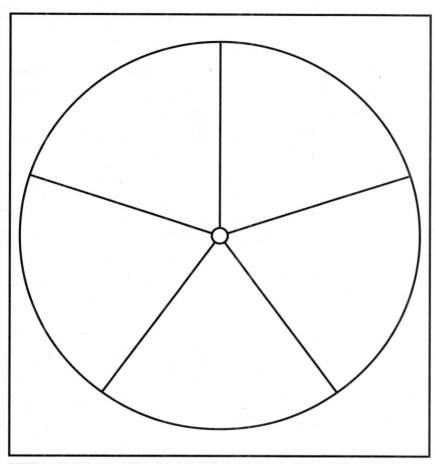

How to assemble spinner.
- Glue patterns to oaktag.
- Cut out and attach pointer with a fastener.

Alternative
- Children can use a paper clip and pencil instead.

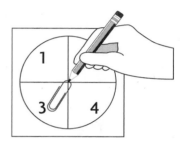

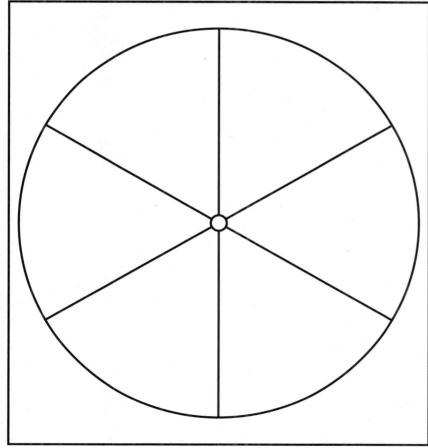

Spinners (5- and 6-section)

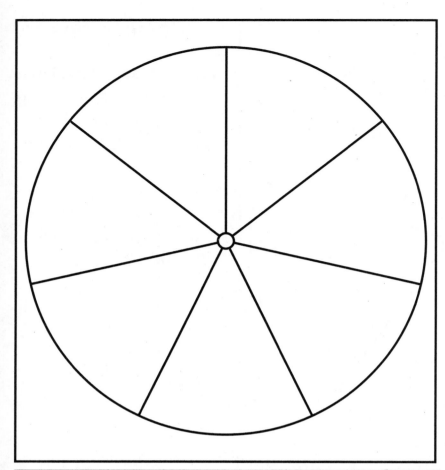

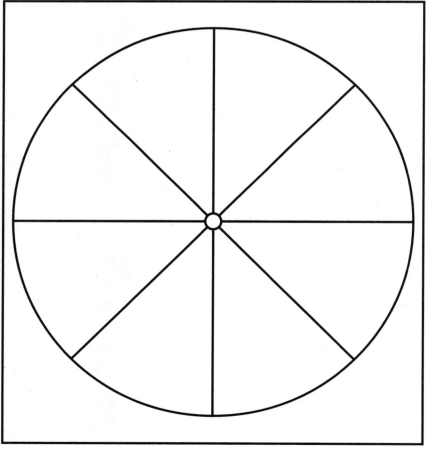

Spinner Tips

How to assemble spinner.
- Glue patterns to oaktag.
- Cut out and attach pointer with a fastener.

Alternative
- Children can use a paper clip and pencil instead.

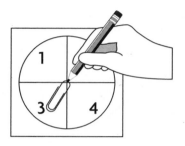

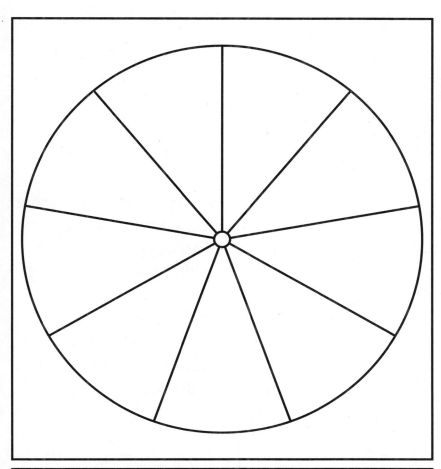

Spinner Tips

How to assemble spinner.
- Glue patterns to oaktag.
- Cut out and attach pointer with a fastener.

Alternative
- Children can use a paper clip and pencil instead.

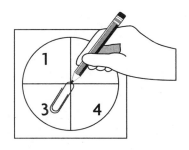

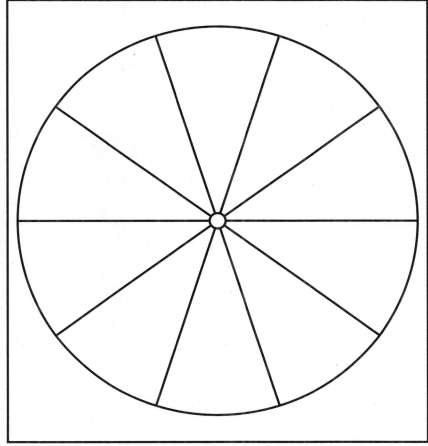

Spinner Tips

How to assemble spinner.
- Glue patterns to oaktag.
- Cut out and attach pointer with a fastener.

Alternative
- Children can use a paper clip and pencil instead.

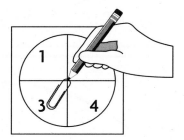

Spinner (unequal sections)

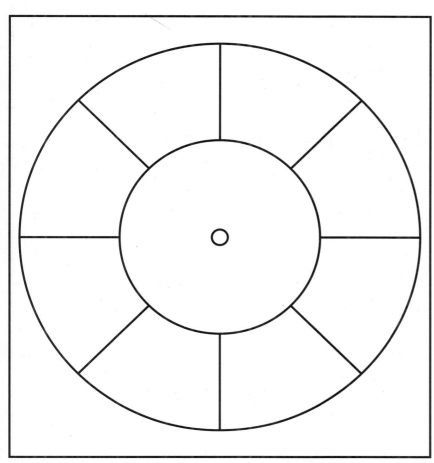

How to assemble spinner.
- Glue patterns to oaktag.
- Cut out and attach pointer with a fastener.

Alternative
- Children can use a paper clip and pencil instead.

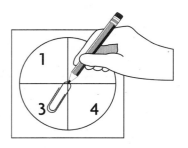

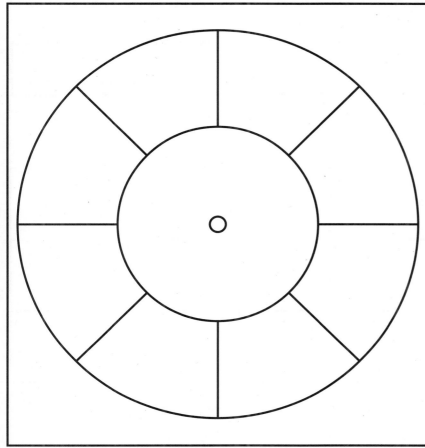

34	35	36	37	38	39	40
33						
32	31	30	29	28	27	26
						25
18	19	20	21	22	23	24
17						
16	15	14	13	12	11	10
						9
2	3	4	5	6	7	8
1						

Hop to It! Gameboard

R99

Workmat 3

Tens	Ones

Workmat 4

Quarter	Dime	Nickel	Penny

Workmat 5

Hundreds	Tens	Ones

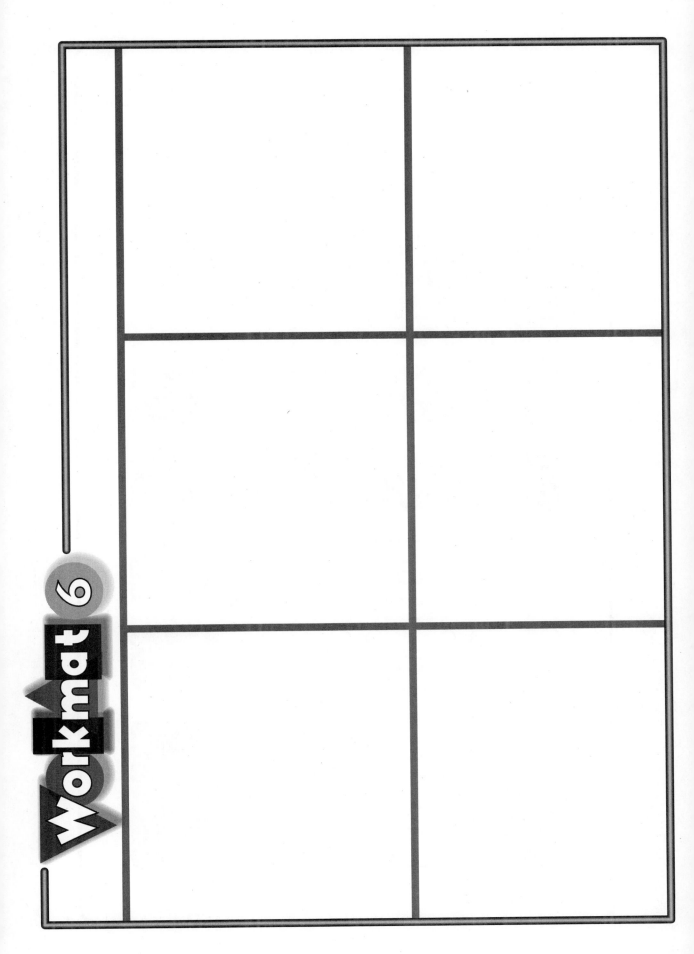

Workmat 6

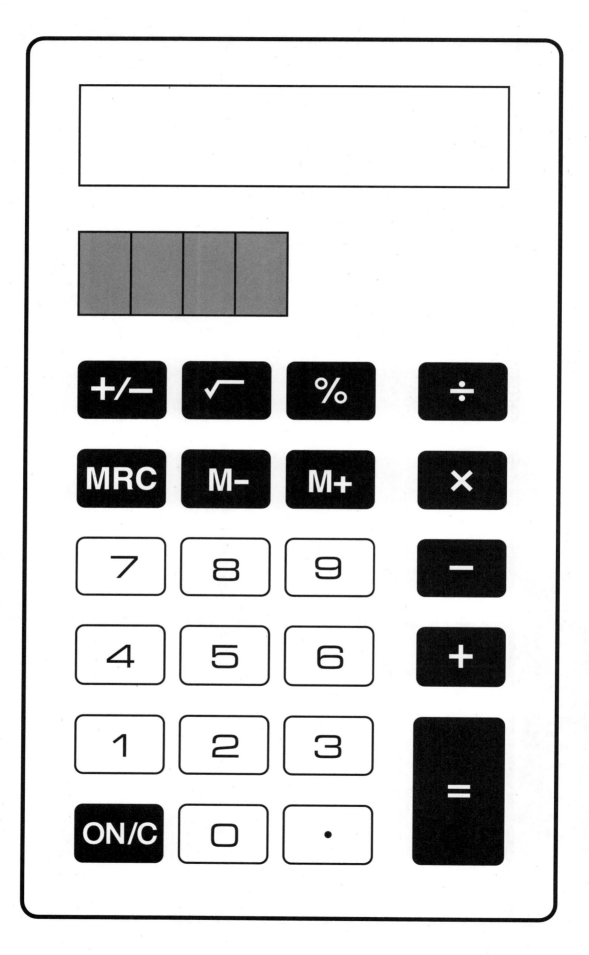

Calculator Keyboard

Standard Keyboard

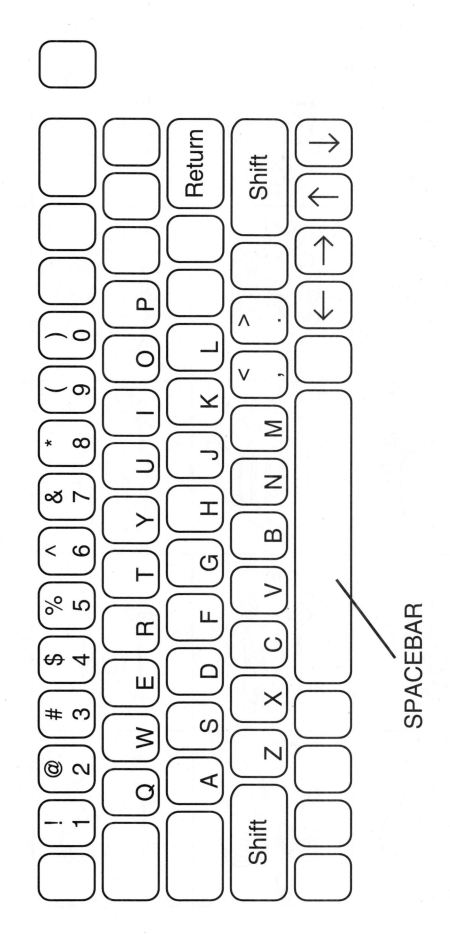

SPACEBAR

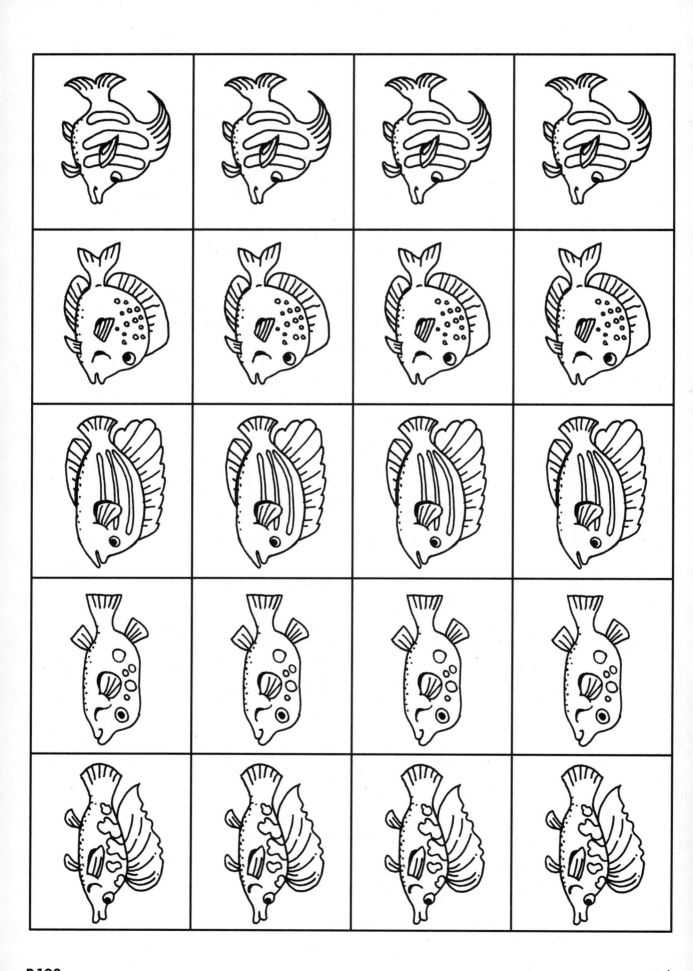

Fish

Fish Bowls

70	66	30
48	9	68
86	51	47

43	61	82
53	17	45
72	55	67

22	47	36
86	34	51
28	61	32

9	53	67
82	45	51
22	70	32

66	30	43
34	72	47
51	28	55

48	22	36
17	68	53
82	32	47

Tic-Tac-Toe Grids

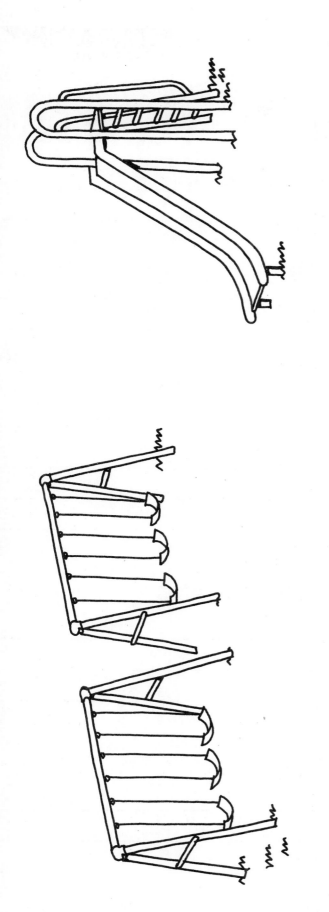
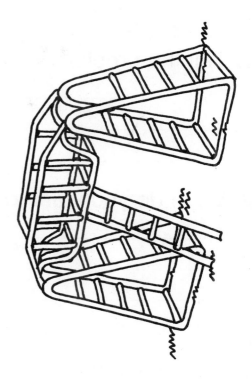

Understand

1. Tell the problem in your own words.

_____

_____

_____

2. What do you want to find out?

_____

_____

_____

Plan

3. How will you solve the problem?

_____

_____

_____

Solve

4. Show how you solved the problem.

_____

_____

_____

Look Back

5. How can you check your answer?

_____

_____

_____

VOCABULARY CARDS

Vocabulary Cards are included in this section. A Vocabulary Card is provided for each vocabulary word that is introduced in this grade level.

There are 4 cards on each page. The front of each card contains the word only. The back of the card contains the word and an illustration or sentence that helps children recall the word and its meaning. The chapter in which the word is introduced is noted on the bottom of each card.

Children can keep their cards in a Math Words File—a container such as a file box, shoe box, or zip-top bag. Have them use their Math Words File to review vocabulary terms and to check spelling. Other suggestions for using the Vocabulary Cards appear in the Teacher's Edition on the Follow-Up Strategies and Activities page at the end of each lesson.

doubles

number
sentence

sum

doubles
plus one

doubles

$4 + 4$
$5 + 5$
$6 + 6$

When you add the same number twice it's called a **double**.

number sentence

$8 - 3 = 5$

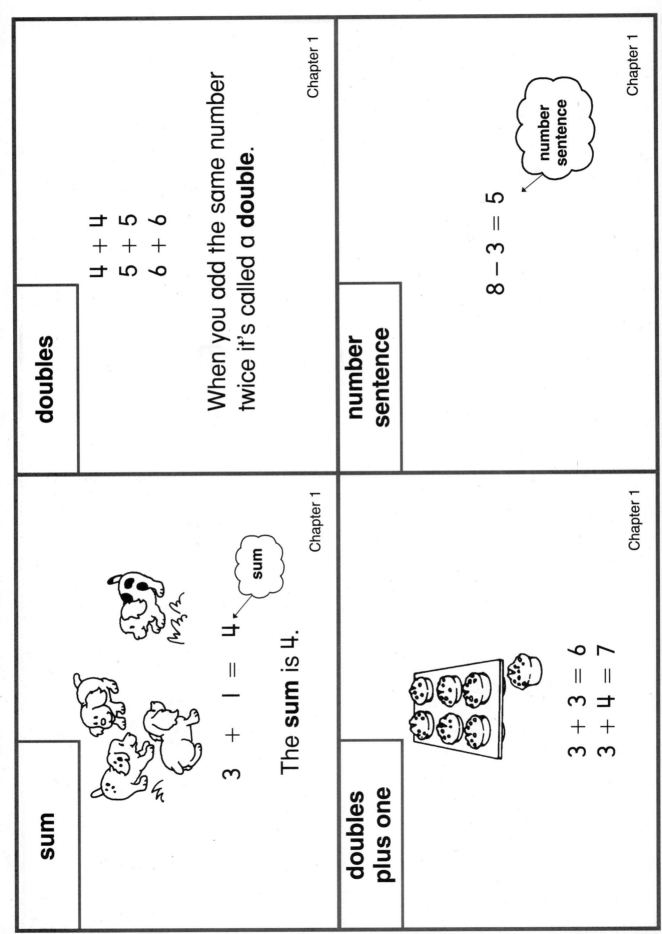

number sentence

sum

$3 + 1 = 4$

sum

The **sum** is 4.

doubles plus one

$3 + 3 = 6$
$3 + 4 = 7$

doubles
minus one

subtract

add

greater

doubles minus one

☆
☆ ☆
☆ ☆
☆ ☆
☆ ☆
☆ ☆

$6 + 6 = 12$
$6 + 5 = 11$

subtract

$4 - 3 = 1$

add

$3 + 2 = 5$

greater

5
● ● ● ● ●

8
● ● ● ● ●
● ● ●

8 is a **greater** number than 5.

addends

number line

count on

fact family

addends

$5 + 2 = 7$

addends

count on

Count on from 5 to add $5 + 3$.

Think 5   Count **6, 7, 8**

$5 + 3 = 8$

number line

0 1 2 3 4 5 6 7 8 9 10

fact family

$5 + 2 = 7$ $2 + 5 = 7$

$7 - 5 = 2$ $7 - 2 = 5$

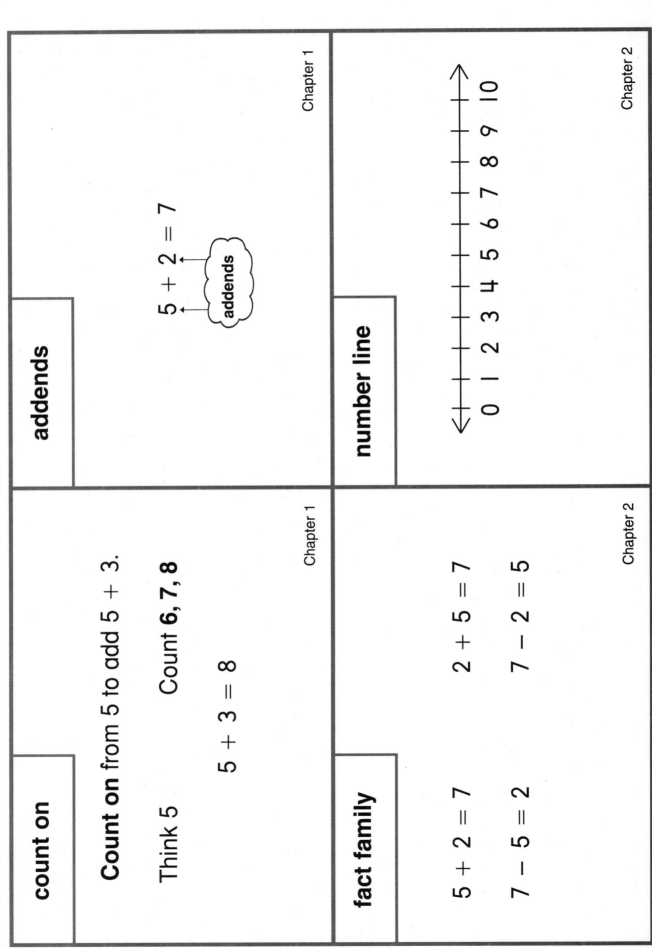

missing
addend

ones

operation

tens

missing addend

missing addend

$$7 + \boxed{8} = 15$$

You need 8 to find the sum.

operation

$$5 + 3 = 8$$
$$6 - 4 = 2$$

Addition is an **operation**.
Subtraction is an **operation**, too.

ones

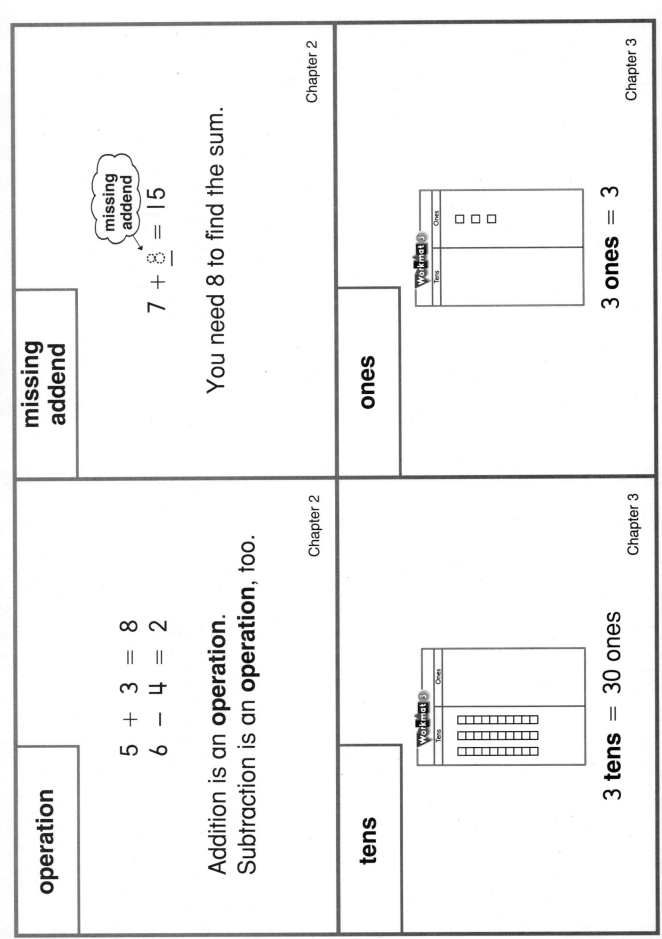

3 ones = 3

tens

3 tens = 30 ones

model

skip-count

estimate

even

model

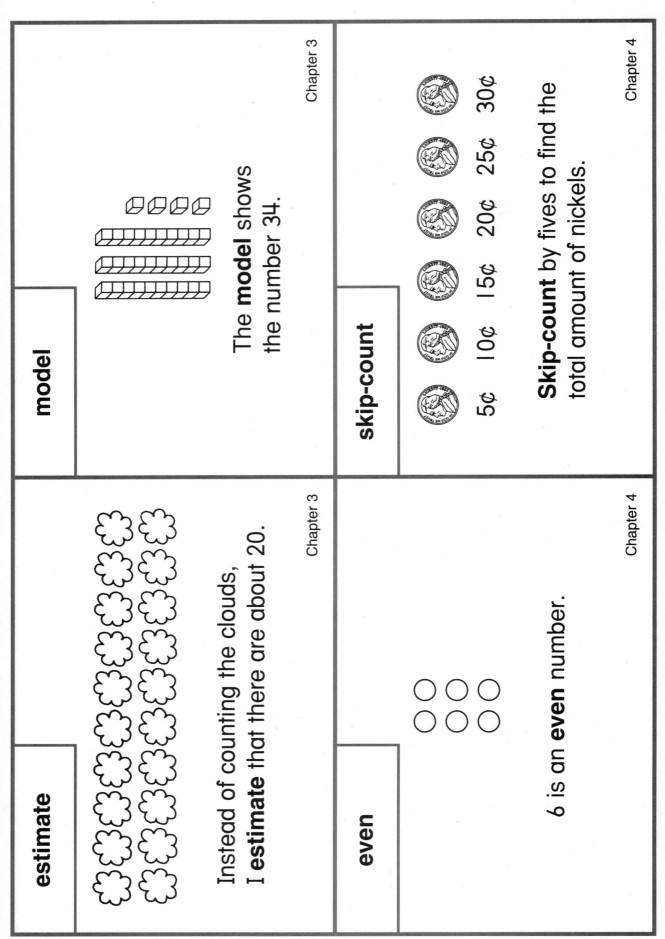

The **model** shows
the number 34.

skip-count

5¢   10¢   15¢   20¢   25¢   30¢

Skip-count by fives to find the
total amount of nickels.

estimate

Instead of counting the clouds,
I **estimate** that there are about 20.

even

6 is an **even** number.

in all

odd

pattern

count back

odd

5 is an **odd** number.

in all

$3 + 2 = 5$

There are 5 turtles **in all**.

count back

$10 - 2 = 8$

10, 9, 8

0 1 2 3 4 5 6 7 8 9 10 11 12

Count back to find $10 - 2$.

pattern

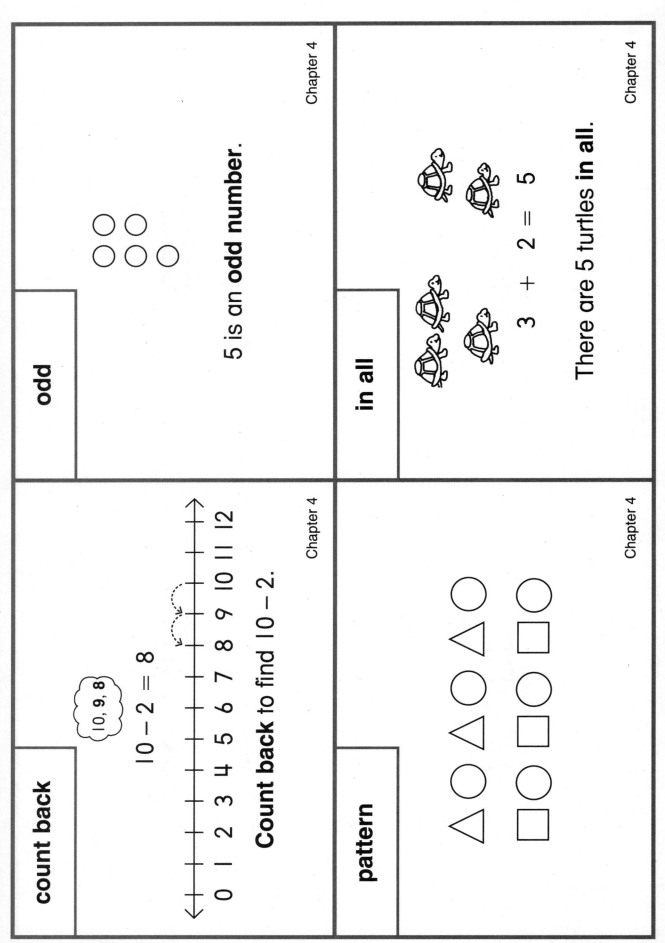

less

rule

greater than

\>

compare

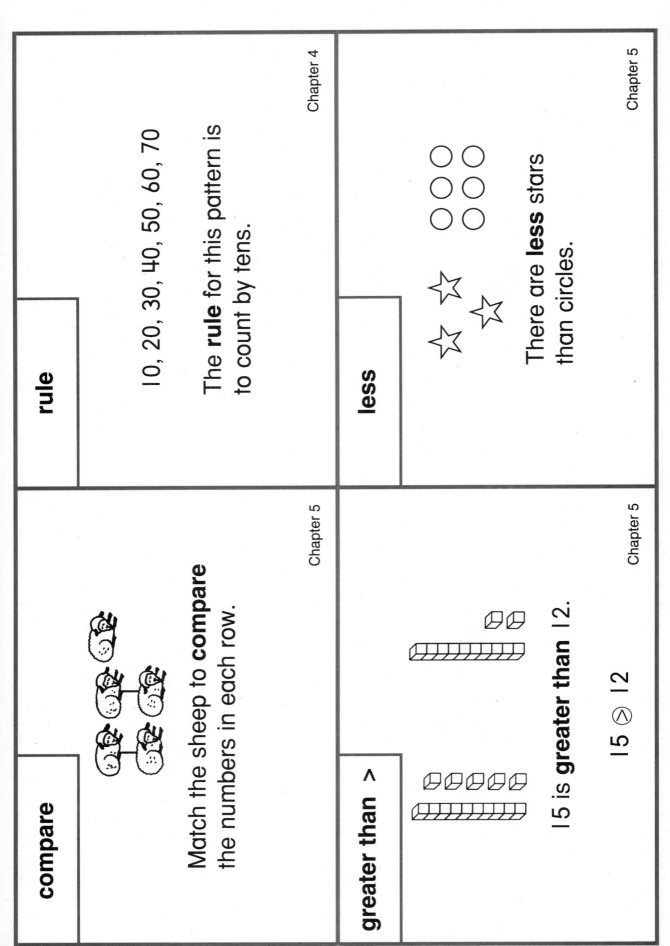

rule

10, 20, 30, 40, 50, 60, 70

The **rule** for this pattern is to count by tens.

less

There are **less** stars than circles.

compare

Match the sheep to **compare** the numbers in each row.

greater than >

15 is **greater than** 12.

15 ⊘ 12

less than

<

before

after

between

less than   <

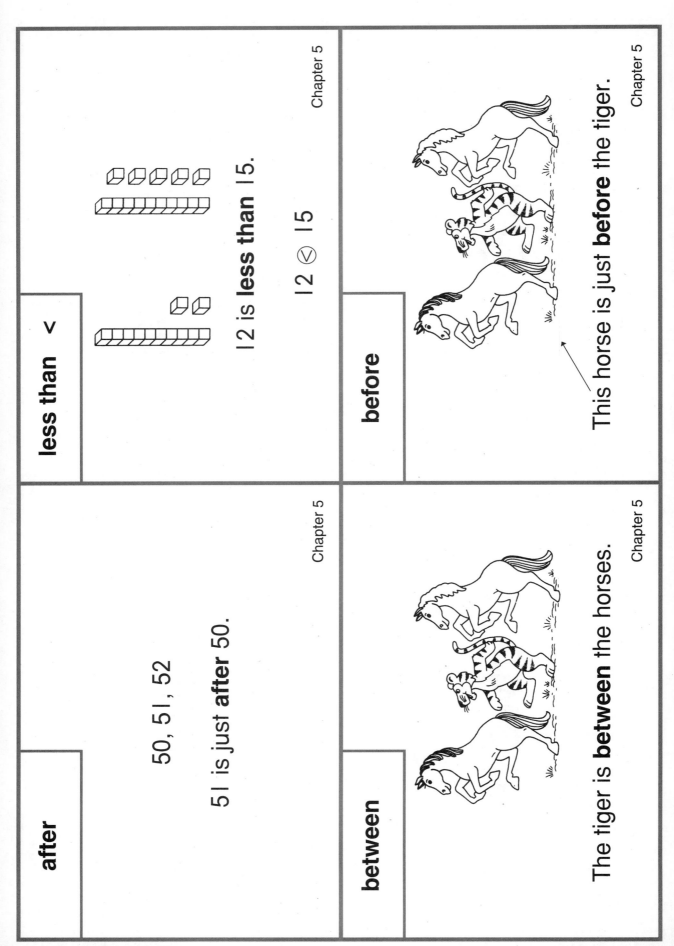

12 is **less than** 15.

12 ⟨⟩ 15

after

50, 51, 52

51 is just **after** 50.

before

This horse is just **before** the tiger.

between

The tiger is **between** the horses.

order

first
1st

ordinal

second
2nd

order

1, 2, 3, 4, 5, 6, 7, 8
These numbers are in **order**.

1, 2, 4, 6, 5, 8, 7
These numbers are <u>not</u> in **order**.

ordinal

The zebra is not one in line.
He is **first** in line.

We use **ordinal** numbers such as first, second, and third to show position.

first (1st)

The mother is the **first** in line.

second (2nd)

The **second** fish is black.

third
3rd

fifth
5th

fourth
4th

sixth
6th

third (3rd)

1st 2nd 3rd 4th 5th 6th 7th 8th 9th 10th

The **third** key is black.

fourth (4th)

The **fourth** bear is circled.

fifth (5th)

1st 2nd 3rd 4th 5th 6th

The **fifth** pig is circled.

sixth (6th)

1st 2nd 3rd 4th 5th 6th 7th

The **sixth** penny is underlined.

seventh
7th

ninth
9th

eighth
8th

tenth
10th

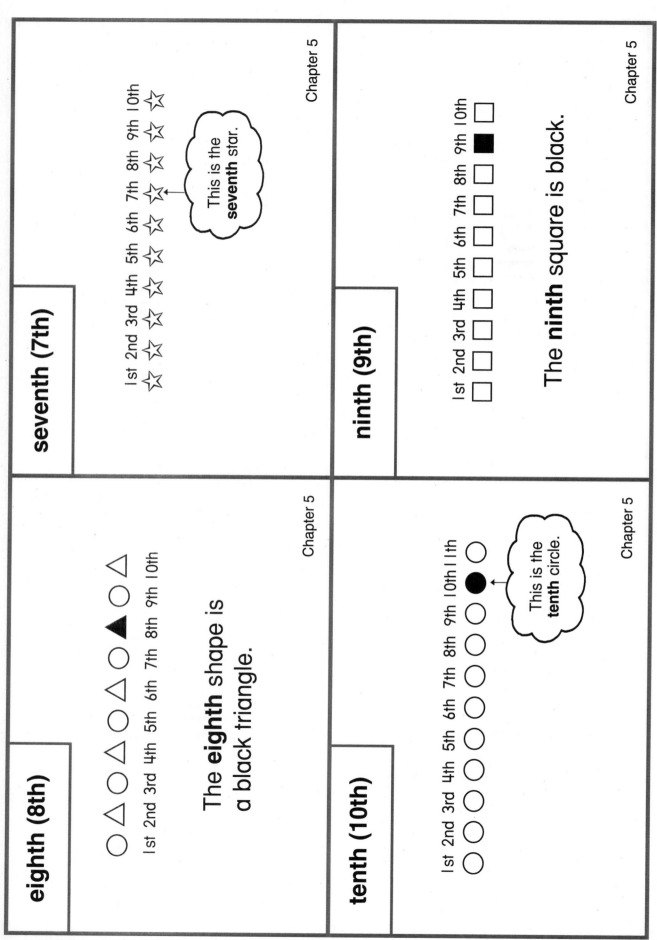

seventh (7th)

1st 2nd 3rd 4th 5th 6th 7th 8th 9th 10th

This is the **seventh** star.

Chapter 5

ninth (9th)

1st 2nd 3rd 4th 5th 6th 7th 8th 9th 10th

The **ninth** square is black.

Chapter 5

eighth (8th)

1st 2nd 3rd 4th 5th 6th 7th 8th 9th 10th

The **eighth** shape is a black triangle.

Chapter 5

tenth (10th)

1st 2nd 3rd 4th 5th 6th 7th 8th 9th 10th 11th

This is the **tenth** circle.

Chapter 5

eleventh 11th	thirteenth 13th
twelfth 12th	fourteenth 14th

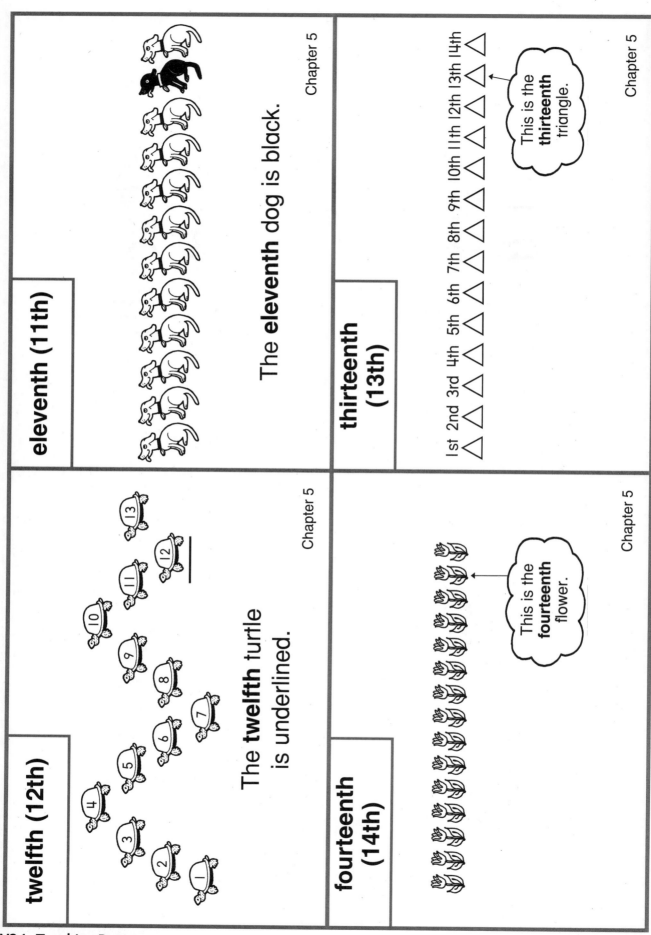

eleventh (11th)

The **eleventh** dog is black.

Chapter 5

twelfth (12th)

The **twelfth** turtle is underlined.

Chapter 5

thirteenth (13th)

1st 2nd 3rd 4th 5th 6th 7th 8th 9th 10th 11th 12th 13th 14th

This is the **thirteenth** triangle.

Chapter 5

fourteenth (14th)

This is the **fourteenth** flower.

Chapter 5

fifteenth
15th

seventeenth
17th

sixteenth
16th

eighteenth
18th

fifteenth (15th)

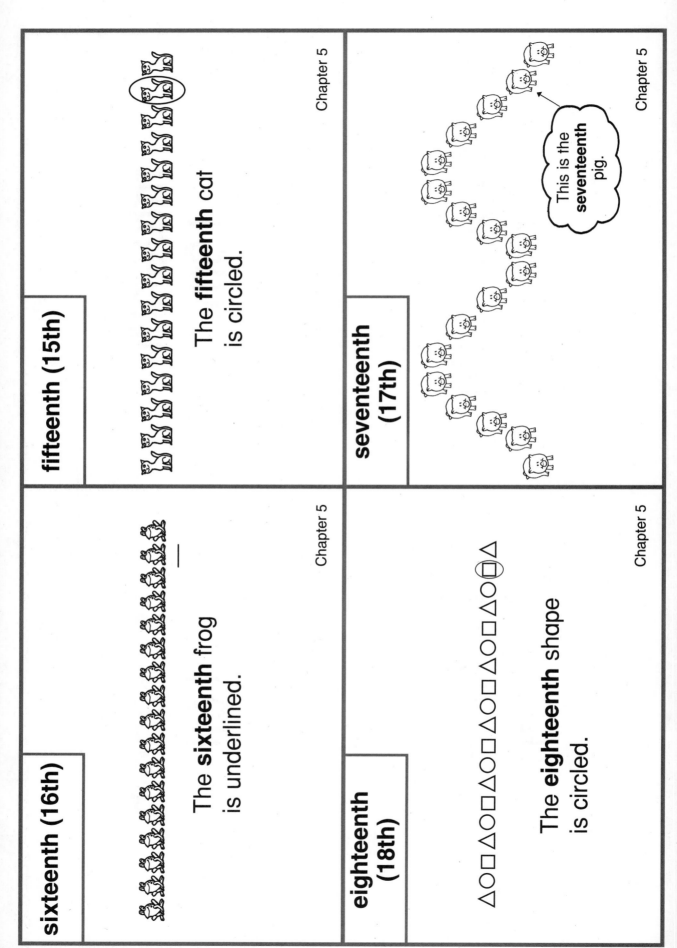

The **fifteenth** cat
is circled.

Chapter 5

sixteenth (16th)

The **sixteenth** frog
is underlined.

Chapter 5

seventeenth (17th)

This is the **seventeenth** pig.

Chapter 5

eighteenth (18th)

△○□△○□△○□△○□△○□△○⬚△

The **eighteenth** shape
is circled.

Chapter 5

nineteenth
19th

penny

twentieth
20th

nickel

nineteenth (19th)

A B C D E F G H I J K L M N O P Q R S T U V W X Y Z

The **nineteenth** letter is an S.

Chapter 5

twentieth (20th)

The **twentieth** footstep got her to the door.

Chapter 5

penny

1¢ or 1 cent

Chapter 6

nickel

5¢ or 5 cents

Chapter 6

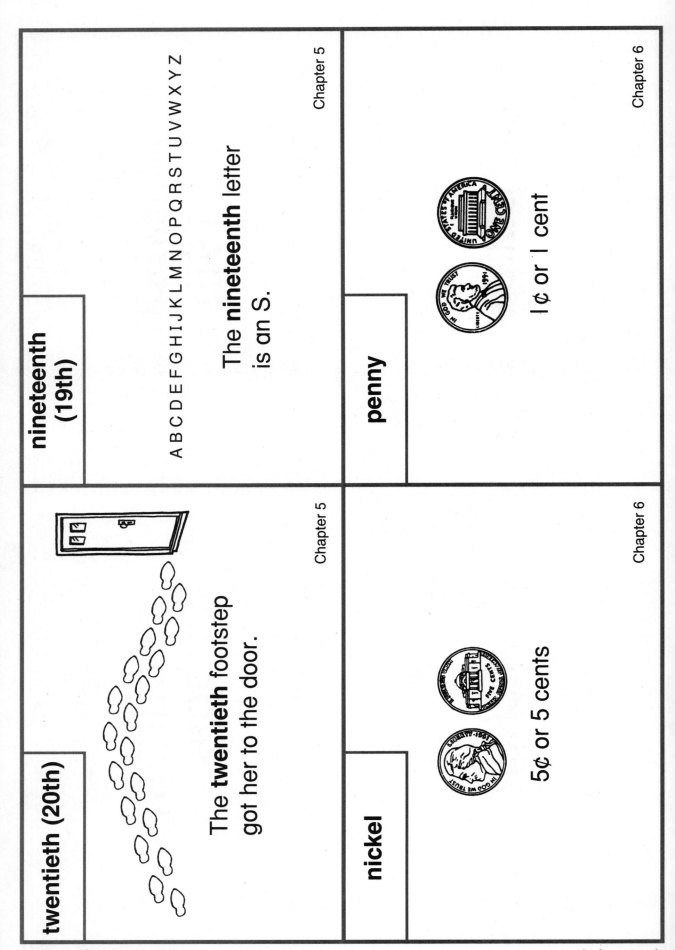

dime

value

total amount

quarter

dime

10¢ or 10 cents

value

25¢ 35¢ 40¢ 41¢ 42¢ 43¢

When you count with coins, start with the coin of greatest **value.**

total amount

The **total amount** is 35¢.

quarter

25¢ or 25 cents

half-dollar

fewer

price

change

half-dollar

half-dollar, 50¢, or 50 cents

fewer

There are **fewer** clouds than there are birds.

price

The **price** of the cookie is 49¢.

change

The candy cost 48¢.
I gave the saleslady 50¢.
I got 2¢ **change**.

half-hour

hour hand

minute hand

minute

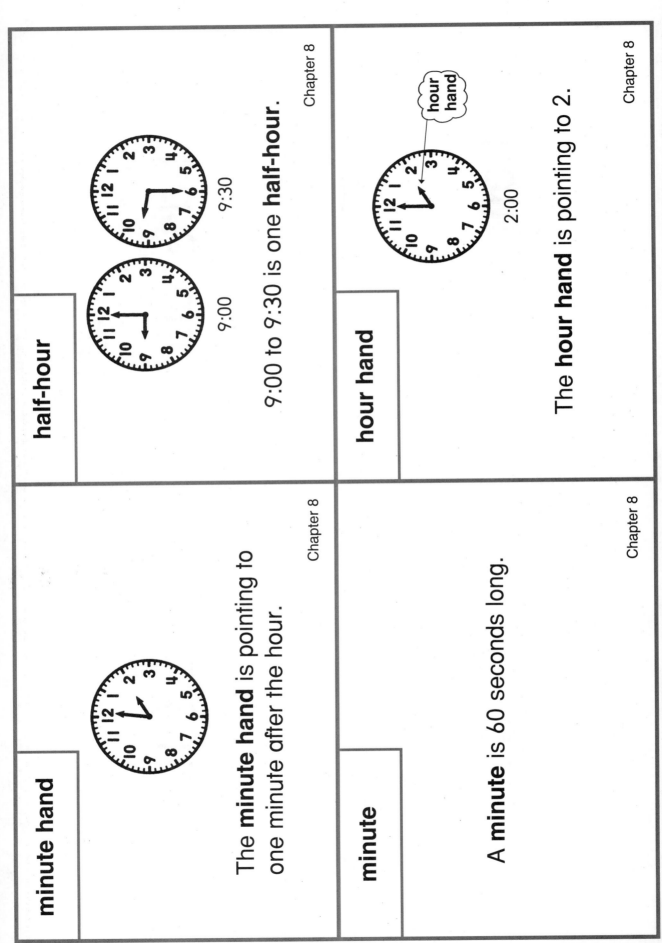

half-hour

9:00

9:30

9:00 to 9:30 is one **half-hour**.

hour hand

hour hand

2:00

The **hour hand** is pointing to 2.

minute hand

The **minute hand** is pointing to one minute after the hour.

minute

A **minute** is 60 seconds long.

calendar

hour

week

month

hour

2:00   3:00

2:00 to 3:00 is one **hour**.

month

January   February
March   April
May   June
July   August
September   October
November   December

There are twelve **months** in a year.

calendar

February

Sunday	Monday	Tuesday	Wednesday	Thursday	Friday	Saturday
		1	2	3	4	5
6	7	8	9	10	11	12
13	14	15	16	17	18	19
20	21	22	23	24	25	26
27	28					

week

Sunday   Monday
Tuesday   Wednesday
Thursday   Friday
Saturday

There are seven days in a **week**.

day

Sunday

date

Monday

day	date

There are 7 **days** is a week.

Three **dates** have been circled.

Sunday	Monday

Sunday is the first day of the week.

Monday is the second day of the week.

Thursday

Tuesday

Friday

Wednesday

Tuesday

Tuesday is the third day of the week.

Thursday

Thursday is the fifth day of the week.

Wednesday

Wednesday is the fourth day of the week.

Friday

Friday is the sixth day of the week.

Saturday

January

year

February

year

There are twelve months in one **year**.

Saturday

Saturday is the seventh day of the week.

February

February is the second month.

January

January is the first month.

May

March

June

April

March

March is the third month.

May

May is the fifth month.

April

April is the fourth month.

June

June is the sixth month.

July

September

August

October

July

July is the seventh month.

September

September is the ninth month.

August

August is the eighth month.

October

October is the tenth month.

November

early

December

late

December

December is the twelfth month.

November

November is the eleventh month.

late

Soccer practice starts at 1:00. Juleen gets there at 1:10. Juleen is **late.**

early

The circus starts at 8:00. Janelle gets there at 7:45. Janelle is **early.**

event

regroup

schedule

join

event

An **event** is an activity that takes place during the day. Math class is an event.

schedule

My Morning

9:00–10:00	Math
10:00–10:30	Recess
10:30–11:30	Language Arts
11:30–12:00	Lunch

A **schedule** is a list showing when events happen.

regroup

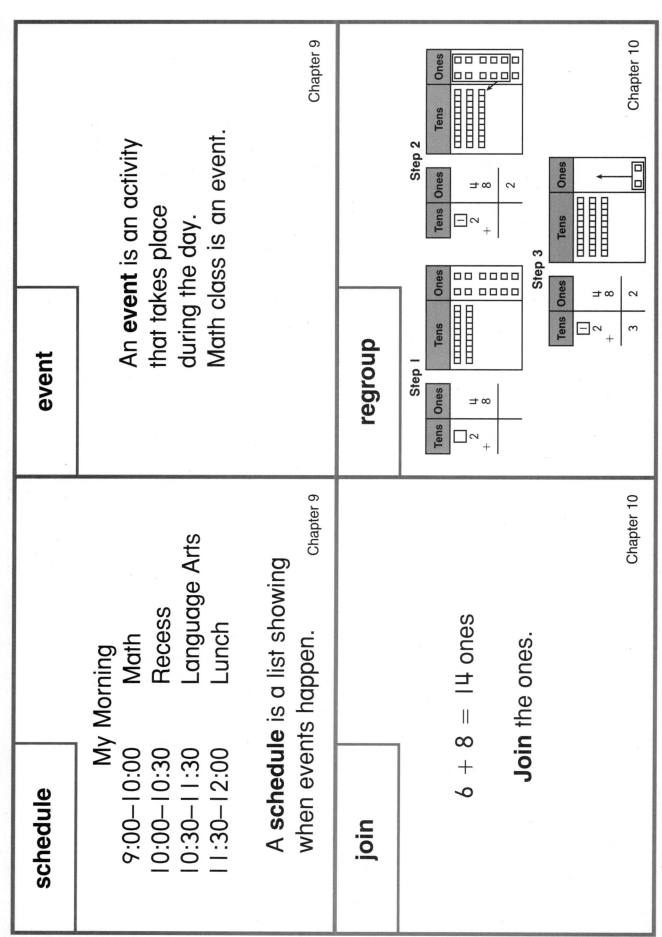

Step 1

Step 2

Step 3

join

$6 + 8 = 14$ ones

Join the ones.

graph

difference

altogether

exact amount

graph

Favorite Pet

	0	1	2	3	4	5
dog						
cat						

We use a **graph** to show data.

difference

4 − 1 = 3

difference

The **difference** is 3.

altogether

Sue had 8 cans.
Mike had 5 cans.
They had 13 cans **altogether**.

exact amount

The **exact amount**
needed to buy
a ball and jump rope
is 68¢.

16¢

52¢

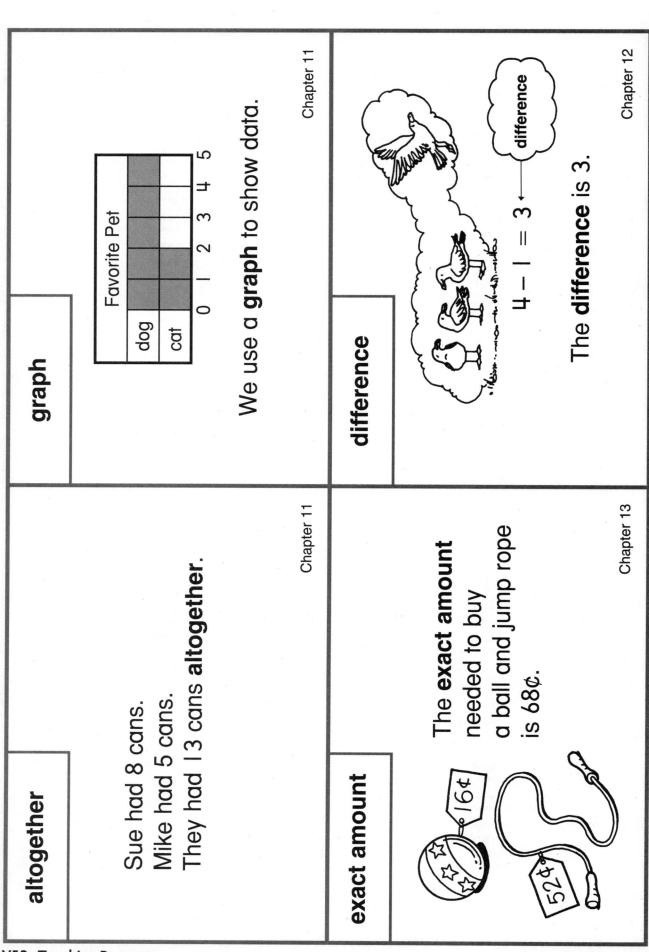

table

tally marks

sort

twice

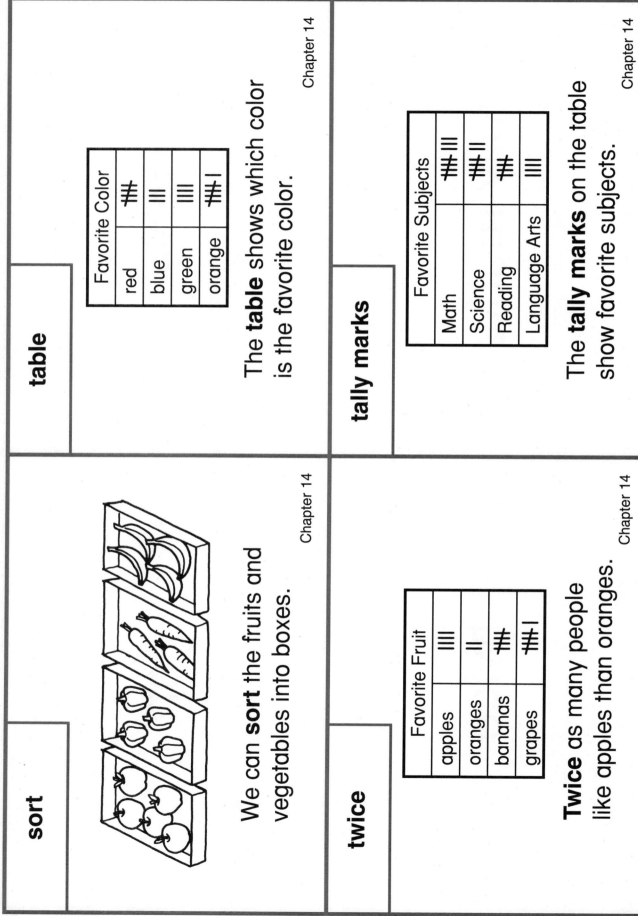

table

Favorite Color	
red	✦ (tally 5)
blue	II
green	IIII
orange	✦I (tally 6)

The **table** shows which color is the favorite color.

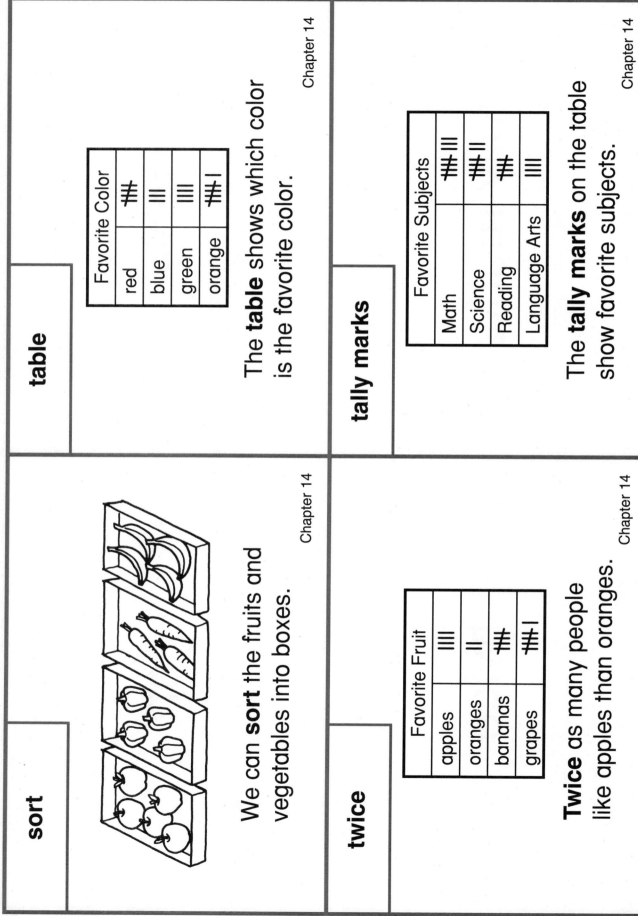

sort

We can **sort** the fruits and vegetables into boxes.

tally marks

Favorite Subjects	
Math	✦ III
Science	✦ II
Reading	✦
Language Arts	IIII

The **tally marks** on the table show favorite subjects.

twice

Favorite Fruit	
apples	IIII
oranges	II
bananas	✦
grapes	✦ I

Twice as many people like apples than oranges.

Chapter 14

survey

data

title

pictograph

survey

Are you a boy or a girl?

When you ask people to answer a question and record the results, you are taking a **survey**.

data

What is your age?	
7	⊞ II
8	⊞ I
9	IIII

A table is filled with **data**, or information.

title

Favorite Fruit	
orange	⊞ ⊞ I
apple	⊞ II
banana	III

The **title** of the table is *Favorite Fruit*.

pictograph

Favorite Activity	
swimming	𝘅 𝘅 𝘅
bike riding	𝘅 𝘅 𝘅 𝘅
playing board games	𝘅

Each 𝘅 stands for 2 children

bar graph

impossible

certain

outcomes

bar graph

Favorite Stories

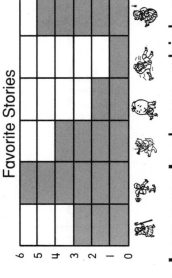

The **bar graph** shows which story
the class liked best.

impossible

It is **impossible** to choose
a triangle from the box.

certain

It is **certain** that a circle, a square,
or a triangle would be chosen
if an object was picked from the box.

outcomes

Shape	Tally Marks
Circle	⊢⊣⊢ lll
Square	ll

The table shows the **outcomes**,
or results, of 10 spins.

least likely

most likely

solid figure

prediction

prediction

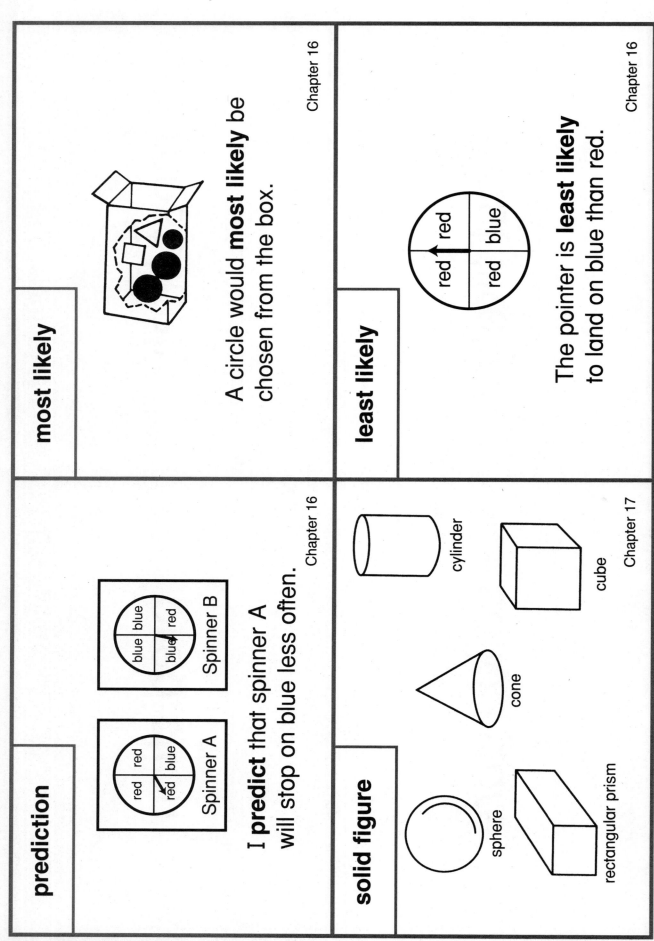

Spinner A

red red
red blue

Spinner B

blue blue
blue red

I **predict** that spinner A
will stop on blue less often.

most likely

A circle would **most likely** be
chosen from the box.

solid figure

cylinder

cube

cone

sphere

rectangular prism

least likely

red red
red blue

The pointer is **least likely**
to land on blue than red.

rectangular prism	sphere
shape	cone

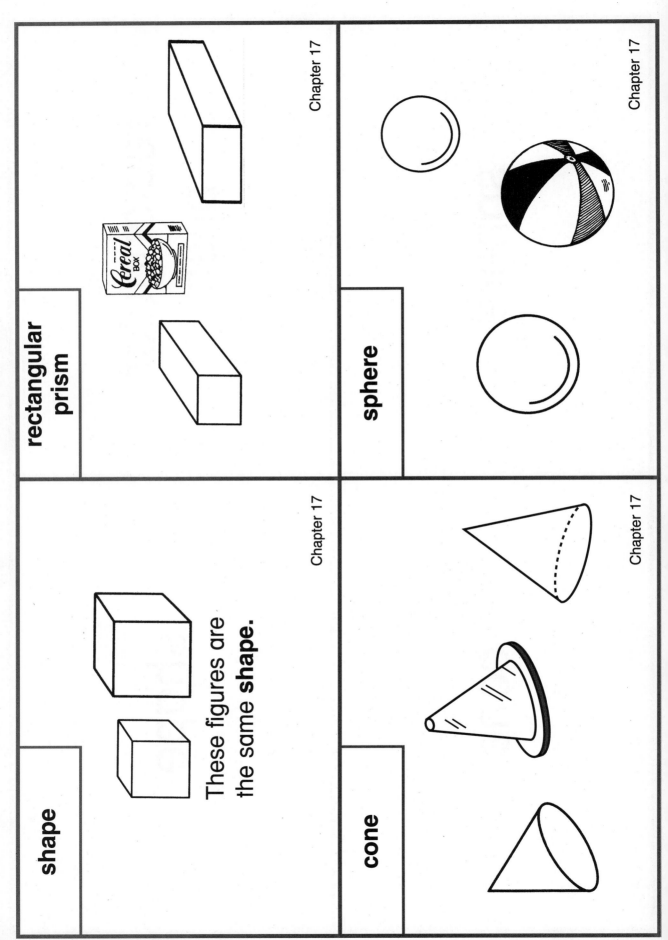

rectangular prism

shape

These figures are the same **shape.**

sphere

cone

pyramid

cylinder

faces

cube

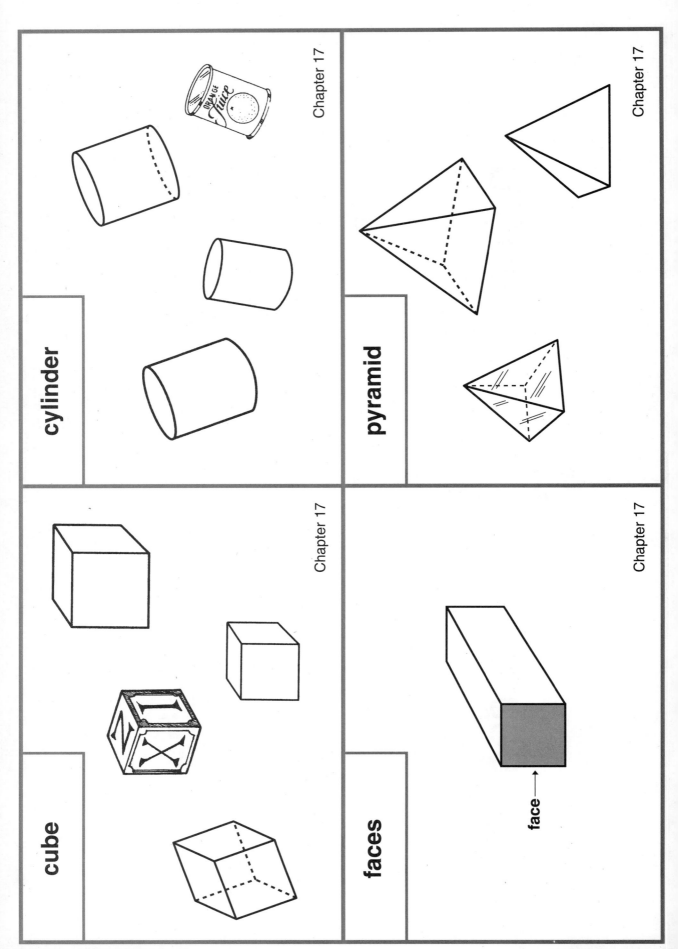

cylinder

pyramid

cube

faces

face

roll

flat

slide

stack

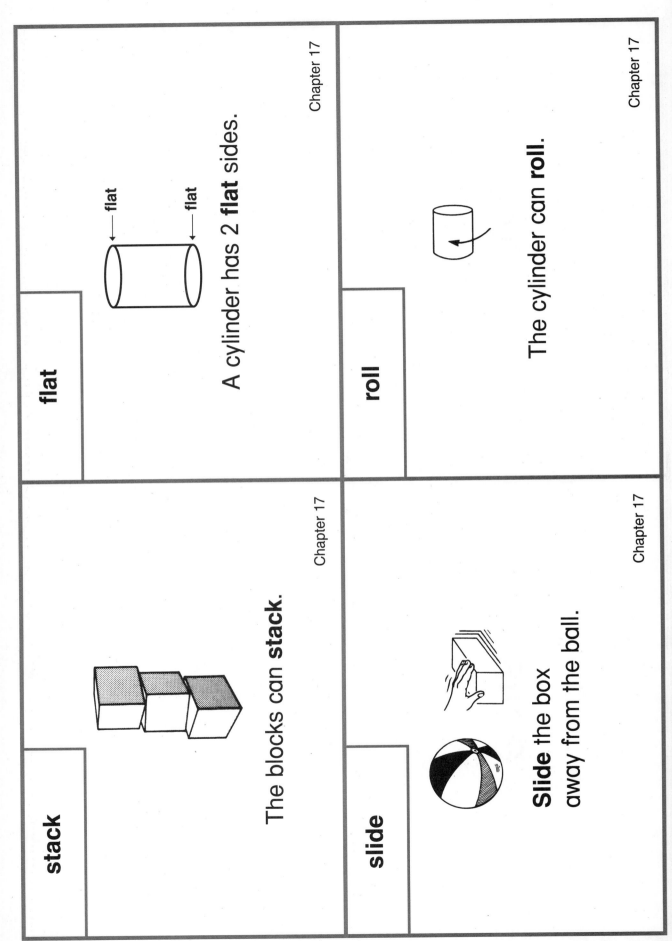

flat

A cylinder has 2 **flat** sides.

flat → flat

Chapter 17

roll

The cylinder can **roll**.

Chapter 17

stack

The blocks can **stack**.

Chapter 17

slide

Slide the box away from the ball.

Chapter 17

| plane figure | square |
| circle | triangle |

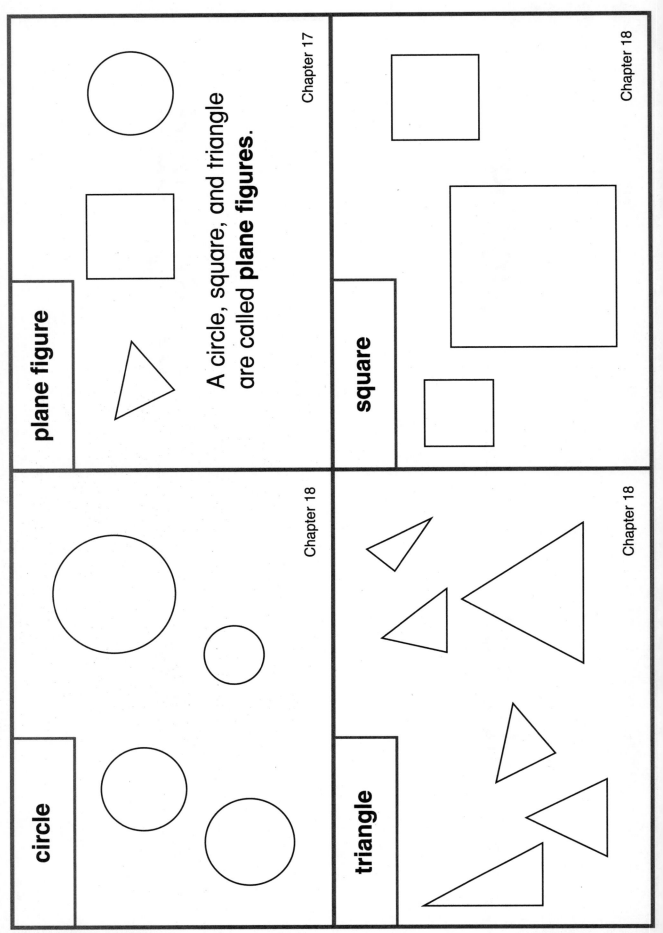

plane figure

A circle, square, and triangle are called **plane figures**.

Chapter 17

square

Chapter 18

circle

Chapter 18

triangle

Chapter 18

rectangle

corner

side

line

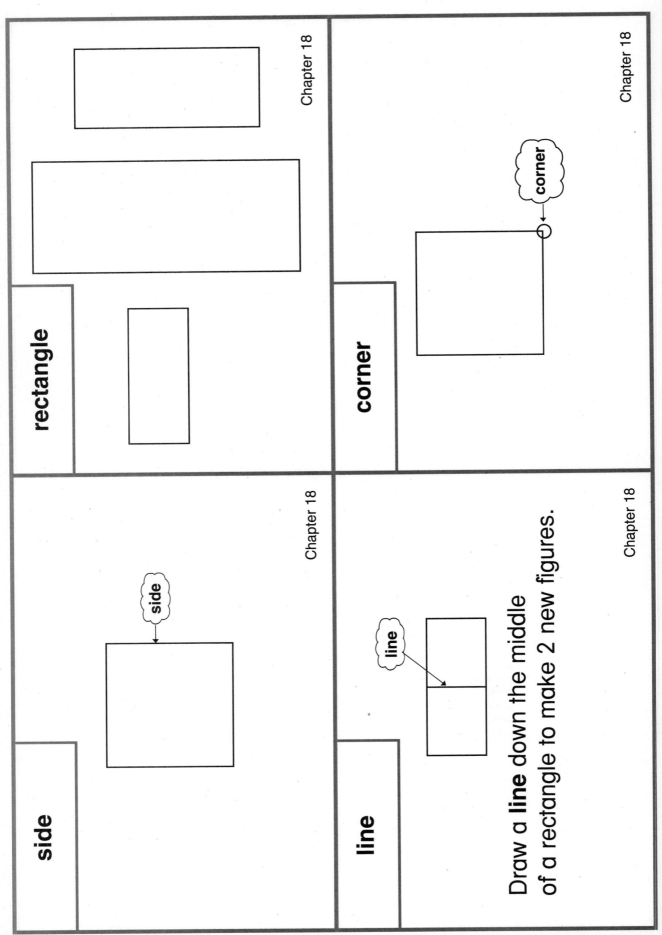

rectangle

corner

corner

side

side

line

line

Draw a **line** down the middle of a rectangle to make 2 new figures.

congruent

line of
symmetry

size

moving
figure

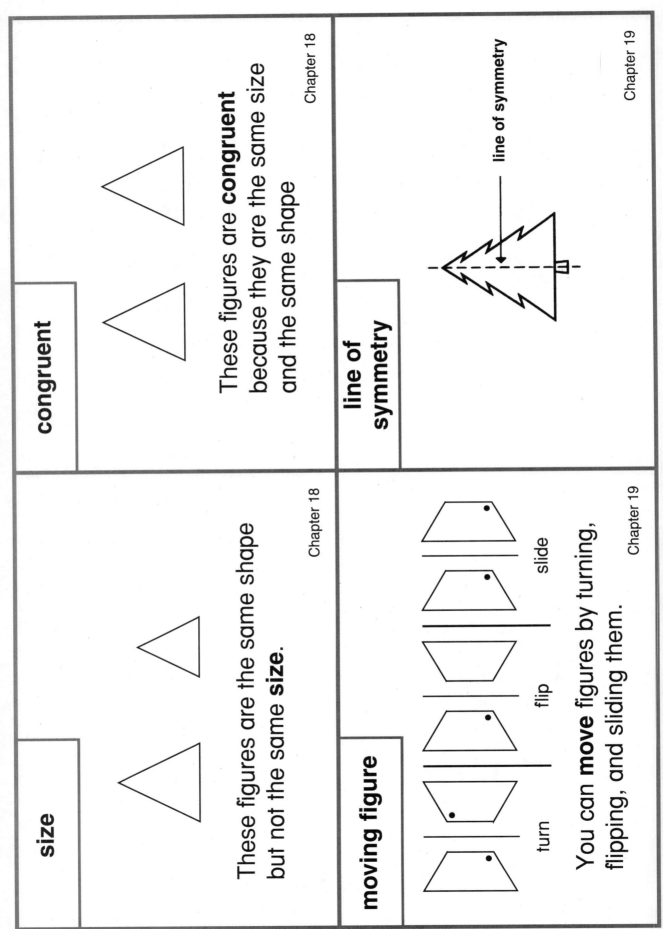

congruent

These figures are **congruent** because they are the same size and the same shape

Chapter 18

size

These figures are the same shape but not the same **size**.

Chapter 18

line of symmetry

line of symmetry

Chapter 19

moving figure

turn   flip   slide

You can **move** figures by turning, flipping, and sliding them.

Chapter 19

measure

flip

long

turn

flip

Pp

We can **flip** the letter.

measure

inches

Use your inch ruler
to **measure** the pencil.

turn

We can **turn** the shape.

long

The yarn is 5 paper clips **long.**

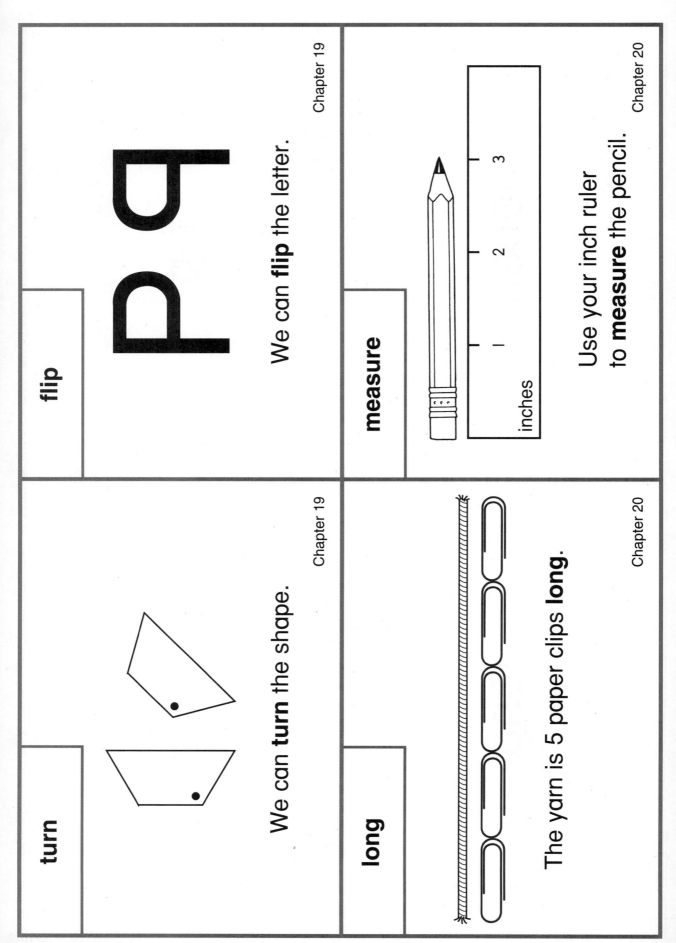

inch

length

ruler

foot

ruler

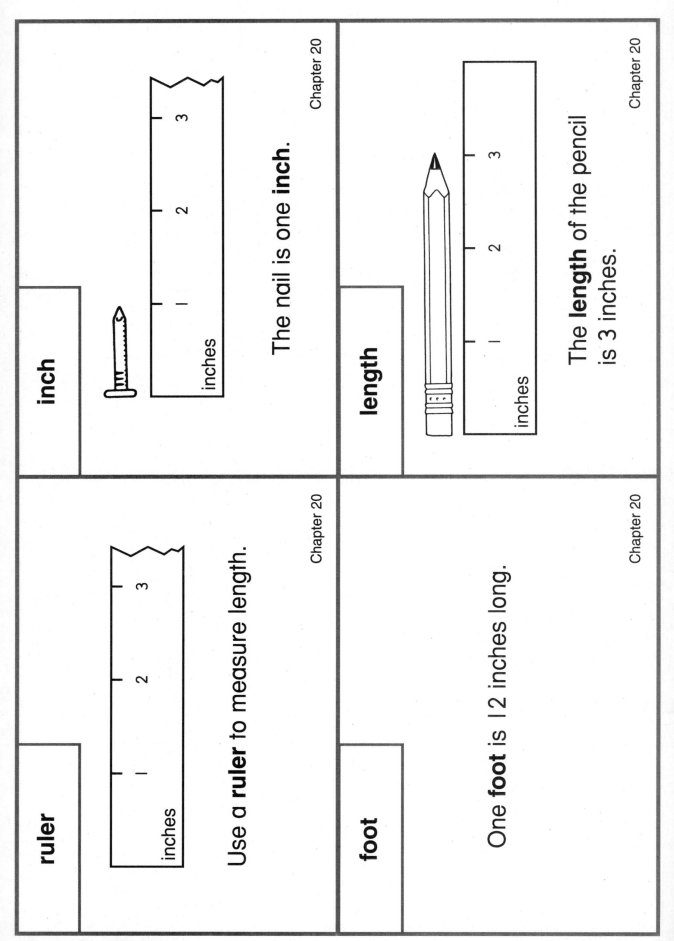

Use a **ruler** to measure length.

inch

inches

The nail is one **inch**.

foot

One **foot** is 12 inches long.

length

inches

The **length** of the pencil is 3 inches.

decimeter

path

perimeter

centimeter

path

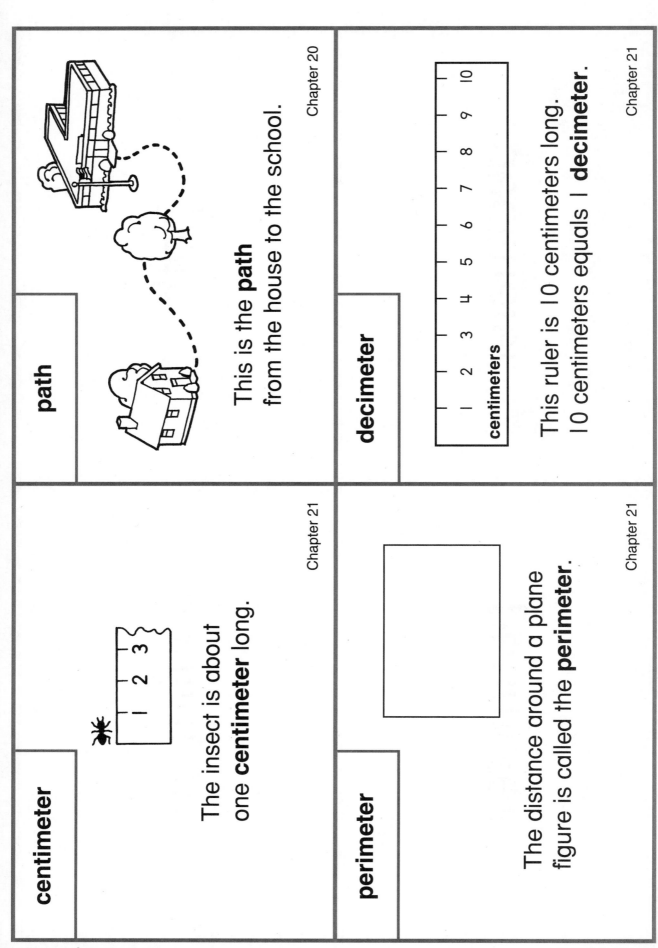

This is the **path** from the house to the school.

decimeter

| | | | | | | | | | |
|1|2|3|4|5|6|7|8|9|10|

centimeters

This ruler is 10 centimeters long. 10 centimeters equals 1 **decimeter**.

centimeter

1 2 3

The insect is about one **centimeter** long.

perimeter

The distance around a plane figure is called the **perimeter**.

quart

cup

pound

pint

cup

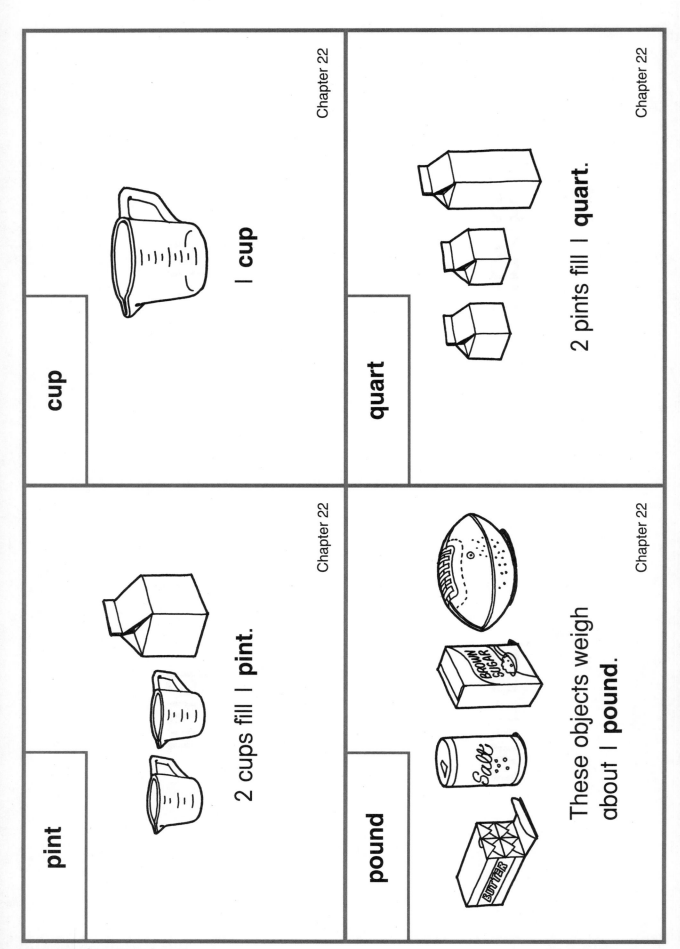

I **cup**

quart

2 pints fill I **quart.**

pint

2 cups fill I **pint.**

pound

These objects weigh about I **pound.**

weigh

temperature

thermometer

degrees

weigh

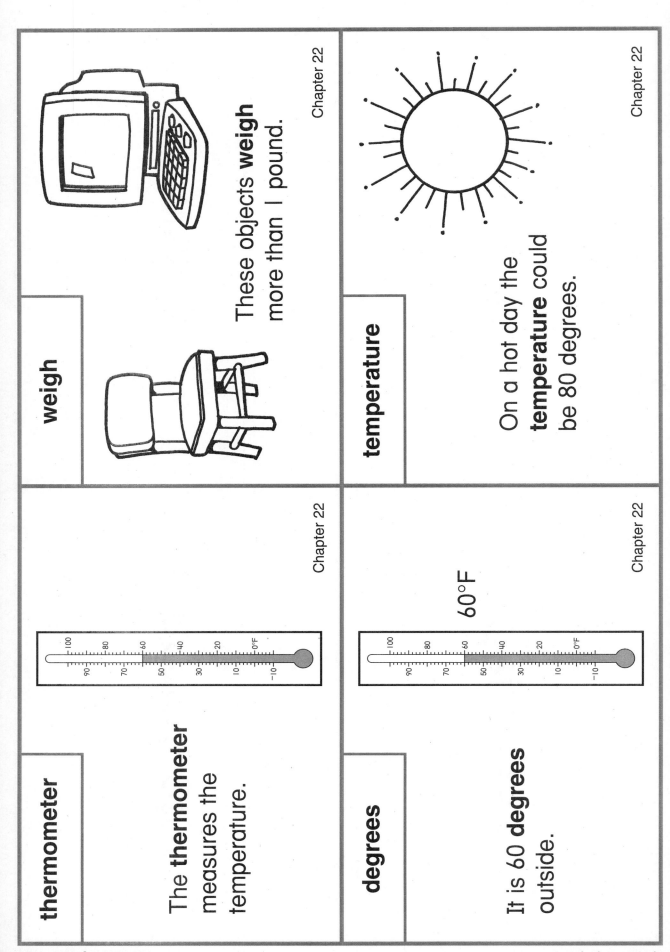

These objects **weigh** more than 1 pound.

temperature

On a hot day the **temperature** could be 80 degrees.

thermometer

The **thermometer** measures the temperature.

degrees

60°F

It is 60 **degrees** outside.

fraction

one part

equal part

halves

fraction

$\frac{1}{2}$ $\frac{1}{3}$ $\frac{1}{4}$

These are **fractions**.

one part

One part of the square is shaded.

equal part

The line shows two **equal parts** of this square.

halves

The sandwich is cut into **halves**.

fourths

one-fourth
$$\frac{1}{4}$$

one-half
$$\frac{1}{2}$$

thirds

fourths

The square has four equal parts.
They are called **fourths**.

one fourth $\frac{1}{4}$

One-fourth of the square is shaded.

one-half $\frac{1}{2}$

One-half of the circle
is shaded.

thirds

The rectangle is divided into **thirds**.
It has 3 equal parts.

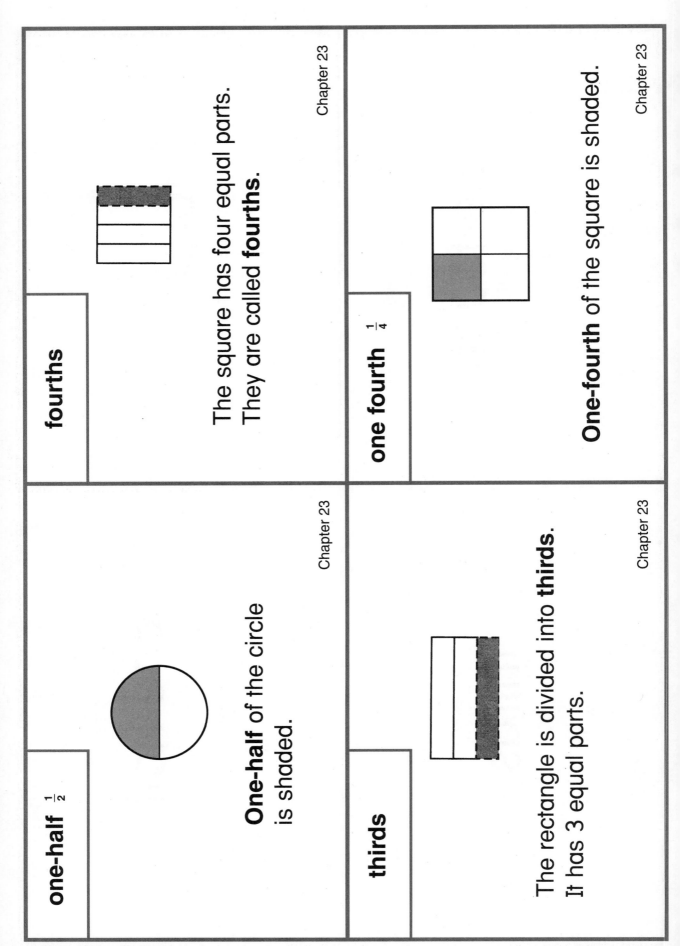

sixths

one-sixth
$\frac{1}{6}$

one-third
$\frac{1}{3}$

two-thirds
$\frac{2}{3}$

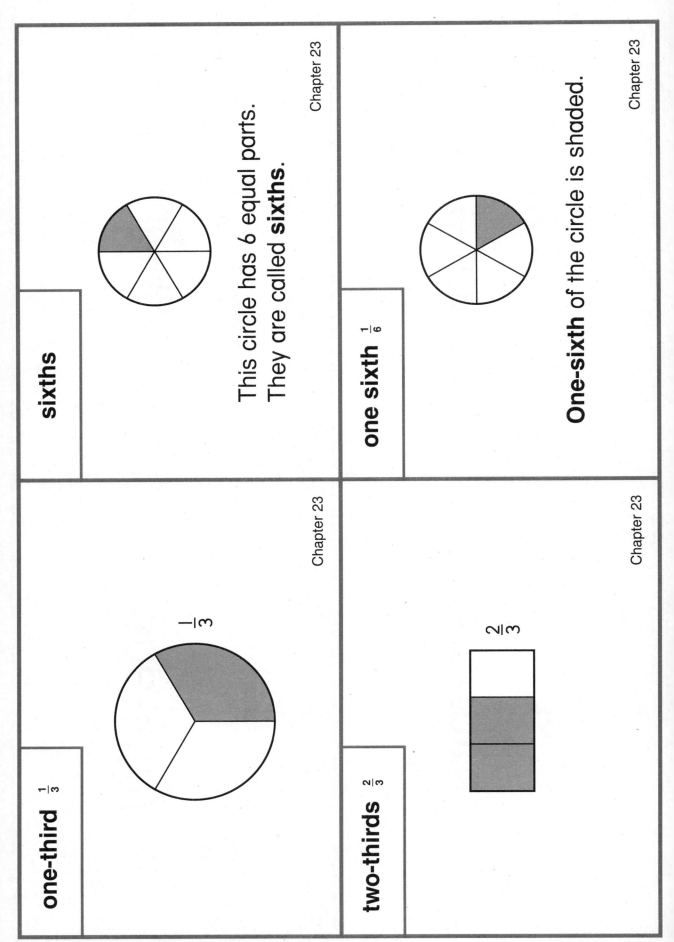

sixths

This circle has *6* equal parts.
They are called **sixths.**

one sixth $\frac{1}{6}$

One-sixth of the circle is shaded.

one-third $\frac{1}{3}$

$\frac{1}{3}$

two-thirds $\frac{2}{3}$

$\frac{2}{3}$

parts of groups

one dollar
$1.00

hundreds

equal groups

parts of groups

There are 2 equal **parts of the group.**

one dollar
$1.00

hundreds

Workmat 5

Hundreds	Tens	Ones

321

There are 3 **hundreds** in 321.

equal groups

There are 3 in each group.
They are **equal groups.**

product

multiplication
sentence

multiply

equal number

product

$5 \times 3 = 15$

15 is the **product**.

multiplication sentence

$2 \times 3 = 6$

multiply

$3 \times 4 = 12$

We **multiply** to find how many.

equal number

The three groups have an **equal number** of fish.